Ministry of Agriculture, Fisheries and Food

Bush Fruits

Reference Book 4
formerly Bulletin 4

London: Her Majesty's Stationery Office

© *Crown copyright 1977*
First published 1977
Second impression (with amendments) 1981

In the interests of factual reporting, occasional reference in this publication to trade names and proprietary products may be inevitable. No endorsement of named products is intended, nor is any criticism implied of similar products which are not mentioned.

Cover photographs: J. Tysterman A.I.I.P.

ISBN 0 11 240527 4

Foreword

This Reference Book deals with black currants, gooseberries and red currants.
In this completely new edition all the information has been brought up to date, particularly on varieties, growing systems, mechanical harvesting and weed control. The Fruit Group of the Agricultural Development and Advisory Service has been generally responsible for this new edition, and the preparation of the text and editing has been undertaken by Miss H M Hughes. The sections on pests were revised by Dr D Alford, diseases by Mr D Wiggell and Mr W T Dale, manuring by Mr D J Eagle and economics by Mr C D Walker. Mr G Shipway prepared the section on mechanical harvesting. The staff of the National Fruit Trials provided the descriptions of varieties, and the ADAS horticultural liaison officer at the Weed Research Organisation the section on herbicides.

R Gardner,
Senior Horticultural Adviser,
Agricultural Development and Advisory Service

Ministry of Agriculture, Fisheries and Food

June, 1976

Contents

	Page
Black Currants	
Growth and Crop	1
production areas	2
improvements in yield	3
disposal of crop	3
Labour Requirements	4
Climate	5
Site	6
spring frost risks—wind frosts	7
radiation frosts—frost pockets and plains	7
slopes—frost assessment of a site	8
Protection against Spring Frosts	8
sprinkler type and spacing	9
operating temperatures	10
water application rate	10
operation of a sprinkler system	11
ancillary uses of solid set sprinkler systems	11
Soil	11
Varieties	13
Pollination	19
Propagation	20
certification schemes—source of cuttings	20
inserting the cuttings—treatment after planting	22
identification of varieties and reversion virus disease	24
Planning the Plantation	25
bush and row spacings	26
multiple row or bed systems	27
plantations direct from cuttings	28
row lengths	29

	Page
Planting	30
soil preparation	30
marking out	31
planting	31
Manuring	33
soil moisture—organic matter	34
soil sampling	35
lime	36
nitrogen	36
phosphorus	36
potassium	37
magnesium	37
minor elements	38
leaf ash analysis	38
Irrigation	39
sprinkler equipment	40
Pruning	41
pruning tools	41
pruning after planting	42
at the end of one year's growth	42
after the first crop	42
after the second crop	42
pruning established bushes	43
pruning bushes for mechanical harvesting	43
disposal of prunings	44
grubbing	44
Control of Weeds	44
principles of chemical weed control	45
contact herbicides	46
residual herbicides	46
translocated herbicides	48
suggested herbicide programmes	48
Spray Applications	50
high volume hydraulic sprayers	52
air-carrier sprayers	53
mixing spray chemicals	53
spray programmes	54
precautions (Appendix II)	106
Main Pests	54
pests of less importance	56
Main Diseases	58
fungus diseases	58
virus diseases	60

	Page
Harvesting	62
hand picking—check point	63
organisation of picking	64
processing trays—market containers	64
mechanical harvesting	65
destructive harvester	65
plucker belt machine	66
vibrating finger machines	67
self-propelled harvesters	67
machine operation	68
machine choice	69
Economics	70
labour and machinery requirements	72
gross margin—budgeting	73
harvesting costs	73

Gooseberries

	Page
Yields and Outputs	77
Site—climate—soil	78
Varieties	80
pollination	81
Propagation	82
form of bush	82
hardwood cuttings	82
Planting	84
preparation of site	84
planting distances	84
Manuring	85
Pruning	87
pruning the young bush	87
pruning the established bush	88
pruning large commercial plantations	88
pruning Leveller	89
Control of Weeds	89
Harvesting	89
Control of Pests and Diseases	90
pests	91
diseases	92
bird and animal damage	95
Economics	96

Page

Red Currants

Growth and Crop	97
Site—soil—climate	97
Varieties	98
Propagation	100
Planting	100
Manuring	101
Pruning and Training	101
Control of Weeds	102
Harvesting	102
Pests and Diseases	102
Economics	103

Appendix I Other Ministry publications 104

Appendix II Precautions 106

Index 109

Illustrations

Plate I (colour) .
    Reverted and healthy flowers and shoots of black currant
    Inspecting black currant bushes for certification

Plate II .
    Growth of black currant bushes four years after planting cuttings, one- and two-year old bushes

Plate III .
    Planting and pruning a black currant bush

Plates IV and V .
    Mechanical harvesters in black currant plantations

Plate VI .
    Pruning an established black currant bush

Plate VII .
    Black currant leaf spot
    Comparison of healthy and reverted black currant leaves

Plate VIII (colour) .
    Modern gooseberry plantations

Line Drawings

Propagation of black currant and gooseberry by cuttings .	23
Pruning a one- and two-year old black currant	24
Pruning a one- and two-year old gooseberry .	93

Black Currants

Growth and crop

The black currant crop is produced on a woody bush generally about $1-1\frac{1}{2}$ m (4–6 ft) high and across. The actual berries are carried on short strigs at the nodes of shoots produced the previous year. These one-year old shoots may be long strong shoots arising from the base (stool) of the bush, or from the older branches, or thin short shoots on the older wood, or tiny short spurs often only a few millimetres long on the two-year old and older branches.

As the shoots age they change colour from the light yellow-brown of the one-year old shoots and become black and thicker. The lower parts of these branches, because they are shaded and crowded in the centre of the bush, become bare of side shoots and spurs.

The flowers of the black currant are initiated during the late summer and autumn in buds in the axils of leaves on shoots of the current year's growth. On the long new shoots some buds at the base may not contain flowers but remain vegetative.

The flower buds may contain only one but generally two or three strigs, each carrying five to fifteen or more flowers. There are marked differences between varieties as to the number of flowers and strigs per bush and season, plantation management and age of bush also affect flower and strig formation. The berries are usually smaller at the end (distal) of the strig and larger at the base where it is attached to the node. At picking the whole strig is removed; berries are generally not picked individually.

During March and April the buds unfold and produce leaves; the flowers unfold and to some extent are protected by the immediate truss leaves. New shoots can be produced from any bud, including those buds that also carry flowers, anywhere in the bush. If the flowers are successfully pollinated the berries start to grow and, depending on variety, are ripe to pick during July and August.

For successful production the black currant bush must cover the area of ground allocated to it as quickly as possible in the early years and thereafter be maintained in a fruitful condition by sufficient removal of older, non-productive wood by pruning. Unpruned bushes become dense and flowers and fruit are only carried on the tips of the shoots. As the size of the individual berry is not important in crops used for processing the aim of the black currant grower is to achieve maximum crop and, unlike all other fruit crops, without the problem of fruit size. Experience shows that once bushes become 10–12 years old it is difficult to maintain a well positioned supply of fruiting wood.

PRODUCTION AREAS

The black currant crop was well established in the fruit growing areas of England and Wales by the early years of the century, mainly as an undercrop to top fruits in Kent and Worcestershire. Since that time the area under the crop has fluctuated widely, reflecting changes in prices and economic conditions and the effects of wartime conditions and policies. A rapid expansion took place after the 1914–18 war, the area rising to a peak of 5400 ha (13 400 ac) in 1929, falling to 3800 ha (9500 ac) in 1934. During the 1939–45 war many old, unfruitful plantations were grubbed and by 1945 the area at 3400 ha (8400 ac) was the lowest since 1913.

Partly due to long term contracts offered by an important black currant juice manufacturer there was an increase to 6500 ha (16 000 ac) by 1950. The improved health of certified bushes being planted and the advent of herbicides in the late 1950s plus the increasing use of irrigation and warm springs resulted in a period of high outputs, 1960–65, and as a result lower prices followed by a reduction in the area grown. In Table 1 the effects of the severe spring frosts in 1967 and 1977 are clearly shown. The main production areas are Hereford and Worcester, Norfolk and Kent.

Table 1 Areas, estimated yields and total output of black currants. England and Wales

	Area (hectares)	Area (Acres)	Yield (tonnes/hectare)	Yield (tons/acre)	Output (tonnes)	Output (tons)
1938–1949 (average)	4035	9971	2·4	·96	9800	9600
1950–1960 (average)	5243	12 956	3·5	1·39	18 600	18 300
1961–1965* (average)	6127	15 140	4·3	1·71	25 400	25 000
1966	4855	11 997	3·5	1·39	17 100	16 800
1967	4473	11 053	1·5	·60	6700	6600
1968	4417	10 915	3·0	1·20	12 700	12 500
1969	4520	11 169	5·4	2·15	24 000	23 600
1970	4206	10 393	4·7	1·87	19 700	19 400
1971	3760	9291	6·0	2·39	22 600	22 200
1972	3738	9237	6·9	2·75	25 300	24 900
1973	3830	9464	5·5	2·19	20 700	20 400
1974	3978	9830	5·7	2·27	22 700	22 300
1975	4092	10 112	5·8	2·31	22 800	22 400
1976	3865	9551	6·5	2·59	17 100	16 700
1977	3739	9239	2·2	0·88	8200	8100
1978	3824	9449	4·6	1·83	17 000	16 700

*Figures up to and including 1963 relate to agricultural holdings. From 1964 figures relate to holdings from which output is commercially significant.

IMPROVEMENTS IN YIELD

For many years the estimated yield was below 2·5 tonne/ha (1 ton/ac). Since 1950 there has been an increasing average yield, except in years when spring frosts damaged the flowers and reduced crop (Table 1). This increase has been due to many causes, chiefly an improvement in the health of bushes, through use of Ministry certified stock for planting; the more careful selection of sites; irrigation; the almost complete replacement of soil cultivation by weed control with herbicides, newer chemicals for effective pest and disease control, and water sprinkling for frost protection.

The effect of the widespread spring frosts of 1967 and 1977 is clearly shown in Table 1. The estimated yield per area is an under calculation since the total area in any one year includes the new, non-bearing plantations, possibly 8–10 per cent of the total, as the black currant does not crop in the first year. The figures are based on field areas, including headlands, as returned by all occupiers of agricultural land.

DISPOSAL OF THE CROP

By far the greatest proportion of the crop is sold for processing in one form or another. At present it is estimated that about 80 per cent of the area is grown for processing, including fixed contracts, and only 20 per cent is offered for sale through the wholesale markets.

Before 1940 the greater proportion of fruit went for jam-making. During the 1939–45 war the value of the high vitamin C content of the fruit was realised and techniques for the manufacture of a vitamin rich juice were evolved and a factory started under government control. After the war the demand for currants for juicing, in addition to the traditional demand for jamming and to a lesser extent for canning, encouraged planting, with the resultant over-production and poor prices in the early 1960s. During this period after the war the wholesale markets received very small supplies of this fruit.

At the present time a large proportion of the tonnage is used for juice manufacture and jamming but demand for these products is fairly static. Demand for canned currants, and for frozen, selected and strigged fruits is increasing, as is that for fruit prepared for pie fillings, yoghourt and the catering trade. A small quantity each year is required by the confectionery trade for use in sweets, pastilles and jellies. In recent years there has been a useful export of fresh and frozen currants, by agents, to the European Community, particularly to Germany and the Netherlands.

It is probably true that the wholesale markets, and hence the housewife, have been under-supplied with black currants. However this fruit is seldom consumed raw and although consumption could no doubt be increased by better supplies and more effective distribution, the high costs of marketing and customer resistance to higher prices means that this method of selling the crop is not certain and over-supply could rapidly reduce returns.

As plantations are generally retained for about ten cropping years growers prefer to seek long-term contracts for the disposal of all, or the greater proportion, of their crop. As these contracts are fewer and more difficult to obtain many growers, particularly non-specialist farmers, or those with sites that have proved unreliable in cropping, have stopped growing this crop. The reduced area has tended to be grown by the specialist grower, or by farmers who have particularly favourable sites.

The commercial introduction of static and mobile mechanical harvesters, particularly two types of mobile harvesters in 1972, may mean that there will be a change in both the localities and units producing this crop over the next decade.

There is only a limited demand for black currants for the pick-your-own trade. Holdings specialising in this method of sale generally have a block of currants with as long a ripening season as possible, obtained through the choice of several varieties.

As with all fruit crops no potential grower should plant, even under the most favourable production conditions, without first ensuring adequate sales outlets. Unless overall consumption can be increased the present output satisfies demand. Any increase in exports demands regular supplies without yield and price fluctuations. There is probably scope for more sales through wholesale markets if marketing costs can be reduced, distribution improved and the season lengthened.

Labour requirements

Once the plantation is established the permanent labour required to grow this crop, including spraying, fertiliser application, pruning and supervision of hand labour is about one man to 6–8 ha (15–20 ac). If pruning is done by casual labour one man with adequate machinery can handle up to 12–16 ha (30–40 ac), but would need help with harvest supervision or mechanical harvesting.

The biggest labour requirement is for picking, where this is to be done by hand. A really skilled hand picker harvesting a good crop from well managed bushes can pick at least 50 kg (110 lb) per normal day, but considerably less than this if unskilled, if the crop is light and needs finding, or if the day is short, or interrupted by rain.

Casual experienced women are becoming more and more difficult to obtain in most areas; this type of labour now seeks regular work throughout the year, and increasingly black currants have to be picked by children, and women who seek outdoor work for short periods. Often this type of labour has to be collected by the grower in his own, or hired, transport, is accompanied by babies or toddlers, needs more supervision and only works a short or broken day. All this adds to overheads, but as black currant harvesting is always done by piece-work, if this type of labour can be obtained and adequately supervised, the harvest can be completed. Similarly, with adequate supervision school children make useful picking labour. About two skilled or four unskilled pickers are needed to pick a tonne (1 ton) of fruit during the harvest season.

Owing to the increasing difficulty in obtaining harvesting labour much effort by research and advisory workers and manufacturers has recently been made to design mechanical harvesters. Several machines are now available (page 65).

In addition to the skilled driver and/or operator, five to eight workers are also required on the machines to handle the empty and full trays and to load and unload the sledges or trailers.

Whether the crop is hand or mechanically harvested the fruit for processing has to be weighed into the wooden trays provided by the processor, loaded and stacked. A tractor-mounted fork lift loader and additional tractors and trailers will be required for moving fruit from the field.

Climate, site and soil

Black currants are a fairly long term crop, costly to establish with no good return on capital until the third or fourth year. The site and soil must be chosen with great care, since no amount of careful crop management or harvesting efficiency can compensate for poor site or soil. The highest yields are obtained from plantations in sheltered, frost-free sites on deep, moisture retentive soils of correct texture.

Climate

Black currants can be grown over a wide range of climates, from Yorkshire in the north, Norfolk and Essex in the east and Kent in the south-east, across southern England to east Devon and to Herefordshire, Worcestershire and Staffordshire in the midlands. They succeed, therefore, under a wide range of rainfall from an annual average of about 630–1000 mm (25–40 in.) or more. Where summer rainfall is low, deep soils with good water reserves are needed and consideration may have to be given to the provision of irrigation, particularly on the lighter soils, but this will increase costs.

The climate in the spring, April-May, is particularly important. Apart from the frost risk there is also the effect of warm days with adequate air moisture (relative humidity) to encourage good leaf production around the early flowers. However very sheltered sites could have the disadvantage of encouraging spring growth too early.

Sites exposed to salt-laden winds near the sea are unsuitable, but experience has shown that well sheltered sites within a few miles of the sea, or estuaries, often yield regularly and above average, due to the lack of spring frosts, the good light conditions, and long growing seasons.

Areas subject to high rainfall, such as parts of Cornwall, Devon and Wales, are unsuitable and give excessive growth and poor flower, and hence crop, production. Areas of a limited growing season, and cool summers such as is experienced on high ground in England and Wales, in north England and in Scotland are unsuitable for currants. With most present varieties the crop is low and the individual berries ripen unevenly, making harvesting difficult.

Site

The selection of a suitable site is the most important factor in growing this crop. Although the deeper soils in valleys may produce larger bushes and heavier crops in seasons free from frosts than those on more elevated sites, the risk of spring frost damage may be too great (page 8). It is better to choose a site as free as possible from spring frost risk, even though the soil may not be quite so good. It is equally important that the bushes should not be exposed to dry cold winds which will be mainly from the east, not only in the spring but throughout the growing season, as these can reduce both growth and crop. Some areas in Kent and East Anglia are subject to this disadvantage. Especially planted windbreaks of suitable subjects such as alder, poplar, birch, pine, ×*Cupressocyparis leylandii* and other quick growing trees can be used to shelter exposed sites. For best results shelter belts, or hedges, need planting several years before the black currants. Artificial windbreaks may be of use to filter and reduce winds at vulnerable places on a plantation.*

Black currants are pollinated by natural pollinators, rather than hive bees, which seldom visit this crop as the bushes are generally in flower at times and temperatures when hive bees are not foraging. Various bumble and solitary bees, and probably smaller insects, are the main pollinators. Plantations sited in areas with plenty of hedges, woods and natural grass fields probably have an advantage as more pollinating insects are likely than in areas under intensive arable or fruit growing conditions. Also insects will forage and fly more readily in sheltered plantations.

Steep slopes, although they may aid air drainage, are difficult to negotiate with machinery and may give rise to erosion problems.

The shape of the site needs study, depending on the method of harvesting. For hand picking very long rows are unsuitable and cross alleys should be provided to reduce walking time to the tally points. With mechanical harvesting the expected average crop per row length, when the plantation is established, should be not more than the capacity of the machine, or extra sledges and trailers will be needed in the middle of the rows, or cross alleys provided. Similarly short rows yielding below machine trailer capacity will considerably reduce machine harvesting efficiency and involve much turning on the headlands.

Access of the site to roads for movement of pickers and to convenient central loading and stacking points for fruit should be considered, together with supply points of water for spraying, or irrigation.

One point worth noting is to avoid sites where trees have been recently felled leaving stumps, with the possible danger of *Armillaria* infection. This soil fungus readily attacks black currants (page 60). If black currants follow apple, pear, plum or cherry orchards there is less likelihood of trouble from this disease, if it was not present in the original plantation, than from grubbed natural woodland. Such orchard sites should be suitable for black currants, provided they are carefully cleared, preferably in autumn under dry conditions, and the soil

---

*Information on the planting and siting of shelter belts is given in the Ministry's Booklet HG 21 *Windbreaks* and in the Fixed Equipment of the Farm Leaflet No. 15 *Shelter Belts for Farmland*.

adequately prepared. The site must not be susceptible to spring frost and will be particularly suitable if existing shelter belts can be retained.

SPRING FROST RISKS

The flowers of black currants are liable to damage from temperatures below minus 1°C (30°F). This damage can occur to the small unopened flower buds at the grape stage (Plate I) as well as the unopened and open flowers and the tiny berries. Thus the period of risk extends from about the last week in March or early April through until late May, when in general all risk of night frost is past.

WIND FROSTS

The frost risk of any place depends on area as well as the site. For example East Anglia is regionally more liable to frost than Somerset, but site factors may act in such a way that an individual site in Somerset is more liable to frost than one in Norfolk.

Two types of frost may occur, wind frosts and radiation frosts. Wind frosts, as the name indicates, are caused by winds in which the air temperature is below freezing; it is not necessary for the winds to be strong and quite light winds with sub-freezing temperatures may cause severe damage.

Wind frosts are infrequent in late spring, but when they occur, they particularly affect plantations in exposed places. Their effects may be lessened but not avoided by the provision of suitable shelter belts. Records from Kent show that in March and April almost half of the frosts which occur along the coast and on hilltops are wind frosts; on hillsides and plains the proportion is about one-quarter.

RADIATION FROSTS

Radiation frosts occur on clear windless nights when all objects, including vegetation, are losing more heat (by radiation) than they are receiving. Their surfaces and the air in contact with them suffer a drop in temperature as a result of such loss. If the cooling continues long enough, the temperature falls below freezing unless the air is sufficiently humid for a fog blanket to form, which may delay or sometimes prevent the formation of frost. If the air is dry and there is little or no wind, the depth of cold air over the level ground slowly increases throughout the night, the coldest air remaining nearest the ground with little or no lateral movement. On sloping ground, this layer of cool dense air flows slowly to lower levels and the areas where cold air collects are those most liable to severe radiation frost. These collecting areas are the so-called 'frost pockets'; they are the hollows and valley bottoms where mist usually first forms. Such areas can often be identified by the behaviour of smoke on still clear nights. This tends to rise vertically in the cold air but to spread horizontally when it meets the warmer air above. These frost pockets can also be formed artificially if the down-slope flow of cold air is impeded by buildings, walls or hedges. The upper part of a moderate slope is relatively free from radiation frost, because the cold air begins to move downhill before any appreciable depth of it has accumulated. The important factor is thus the conformation of the site in relation to the surrounding country.

FROST POCKETS AND PLAINS

Frost pockets may occur in depressions at any altitude, but are most common on valley bottoms. Such areas are sometimes formed naturally and sometimes artificially. Little can be done to improve natural frost pockets but sometimes an artificial one can be remedied by removing the obstruction to cold air drainage. When no improvement can be made, such pockets should be avoided as sites for planting.

Frost plains are broad flat areas such as parts of the Weald of Kent, the Worcestershire plain and the plain of Somerset. During long and severe frosts, the air cooled on the floor of the plain is reinforced by cold air flowing in from surrounding high land and all the fruit on the lower levels is likely to be affected.

SLOPES

A site on a hillside is not likely to be affected by the accumulation of cold air unless boundary hedges have created an artificial frost pocket. It is important to visualise the flow of cold air down the slopes on radiation nights. Its course and depth can often be followed after a frosty night by observing damage to existing shrubs and trees. The lower leaves of susceptible trees such as ash, beech, and oak may exhibit damage up to a certain height, above which they are unhurt; such indications are of great value when choosing fruit planting sites.

FROST ASSESSMENT OF A SITE

Even within a proposed planting area, there are likely to be some parts of the farm more liable to frost than others because of local factors, and it is essential that these areas be recognised. This may be done by exposing a number of accurate minimum thermometers over the area, at places ranging from the highest to the lowest. The thermometers (shielded from radiation) must be read and reset each morning from late March to the end of May and, if planning time permits, this should be done in two consecutive years.

It needs expert interpretation and analysis to use these readings to obtain forecasts of the long term frost frequencies and in some cases this may be available through the Agricultural Development and Advisory Service. However, a detailed examination of the readings will show the range of minimum temperatures that occur on those still clear nights when frost is most likely, and will show the differences in temperature that can occur over a grower's land. It will be realised that this method of recording will not give any indication of the actual duration of frosts.

Protection against spring frosts

Various attempts have been made to protect black currant bushes during periods of falling temperatures and risk of frost damage to the flower trusses and young fruits. In the past these included direct air heating using various forms of heaters burning oil or petroleum products of one form or another. In view of the present high price of oil none of these methods are economic.

When water freezes it releases latent heat and this method of protecting the flowers by water sprinkling has commercial application and is used in some plantations. Sprinkler standpipes arising from overland mains situated down the bush rows can be used to apply water thoroughly to wet the flowers and leaves of the black currants and so protect them from frost. As considerable quantities of water have to be applied on each night that frost is experienced sprinkler systems can only be used on free draining soils, or on slight slopes. On many farms water restrictions apply and the installation of a sprinkler frost protection system may also involve provision of a reservoir, in which the drainage and run off water can be collected.

The frost protection system can also be used later in the year to apply water to irrigate the bushes as required.

Although continuous sprinkling can give a 5°C (9°F) temperature lift, if the method is properly operated, it must be realised that on some sites temperatures lower than this could occur and under these conditions, in spite of sprinkling, frost damage would occur. Thus the provision of sprinkler systems over bushes in frost pockets liable to very low temperatures is likely to be unsatisfactory. Sprinkler systems are of most value on frost plains or slopes where some frost hazard is likely, where water supply problems do not occur, where the soil is not likely to become waterlogged and where light spring frosts occur but other conditions for black currants are good. Under these conditions the high capital cost of the equipment may be justified.

Continuous water sprinkler systems can be effectively used against radiation frosts but are unlikely to give adequate protection from wind frosts as higher quantities of water than those normally applied are necessary, while the wind can also severely distort the sprinkler patterns. Increased damage can result from sprinkling under severe wind frost conditions as the flowers once wet are more susceptible to frost. The protective effect of continuous overhead sprinkling depends on the transfer of heat from the added water, mainly by the release of latent heat on freezing of the water as it turns to ice. This provides approximately 80 calories per gramme of water frozen. In order to achieve protection a film of water must be maintained on the plant throughout the period of frost. As long as there is free water available to freeze, the temperature of the ice and water mixture and the plant remains near to 0°C (32°F), which is above the critical temperature. However, if sprinkling is interrupted for more than a few minutes once the bushes or trees are wet, all the water may freeze and the flowers are then likely to be more damaged than they would be if no sprinkling has been done at all.

SPRINKLER TYPE AND SPACING

Choice of the correct type of sprinkler and the sprinkler spacing and operating pressure are most important. The spring assembly on each sprinkler should be correctly tensioned to give one rotation every 30–60 seconds in order to achieve adequate rewetting of the crop. The spring should also be enclosed to prevent the moving parts from icing up. With the single nozzle sprinklers used for frost sprinkling, the inherent lack of balance necessitates that the sprinkler standpipes are firmly supported to prevent whip of the upright pipe, which can affect speed of rotation and range as well as tending to loosen the screwed connection of the standpipe.

Fairly high operating pressures at the sprinkler nozzle are required with the present type of rotary sprinkler to achieve adequate range and to break up the droplets. In order to obtain as uniform an application of water as possible triangular spacing of sprinklers is usually adopted. To reduce the costs of installation there has been a tendency to increase sprinkler spacings and at extended spacings an equilateral triangle arrangement will usually ensure a better distribution and coverage of water than the more usual isosceles triangle arrangement. The actual spacings will depend on the sprinkler type and nozzle size and row distances but the equilateral principle still applies; generally 17–19 sprinklers will be required per hectare (7–8 sprinklers/ac).

OPERATING TEMPERATURES

The sensing thermometer should be placed and exposed at plant level at a point equivalent to the coldest part of the area to be protected but in a position where it will not be damaged by farm equipment or in other ways. It should be shaded from morning sunlight. Some allowance may be made for evaporative cooling, which occurs under relatively low humidity or wind conditions, by using a wet bulb thermometer to determine the beginning of sprinkling only. A wet bulb thermometer generally gives a lower reading than a dry bulb instrument.

The black currant blossom is particularly difficult to wet, and sprinkling should start when the falling air temperature reaches 0°C (32°F) and should continue until the air temperature remains above 0°C (32°F) and the ice starts melting on its own. The addition of a wetting agent during the first few minutes of sprinkling may ensure more rapid wetting of the blossom.

An adjustable contact mercury thermometer wired through a relay can be arranged to ring a bell in the control room or operator's bedroom (the alarm thermometer being set a little above freezing to allow time for the operator to reach the pump unit), or can be used to set the sprinkler system in operation automatically. Two such adjustable thermometers, together with a suitable relay, can be used to provide a differential on/off temperature operation. Warning equipment can also now be obtained to show at the flick of a switch in the control room or bedroom the temperature in the plantation.

The Meteorological Office provide a frost warning scheme for fruit growers from the London Weather Centre. Warnings are issued by telephone at any time of the day or night when frosts are likely to occur, although with normal radiation frost, they can usually be issued by about 5 pm. Details of current charges can be obtained from the Meteorological Office, London Road, Bracknell, Berkshire.

WATER APPLICATION RATE

As air temperatures in the crop being protected decrease, in theory, water application rates should increase. In practice it is difficult to vary the rate and present recommendations are based on a water application rate of at least 3 mm/hour ($\frac{1}{8}$ in./hour) which will provide a 5°C (9°F) temperature lift. With black currants a higher rate of 4 mm/hour ($\frac{1}{6}$ in./hour) is desirable. Application rates that are too low can increase frost damage. If 3 mm/hour ($\frac{1}{8}$ in./hour) is to be applied a water flow rate of 528 litre/minute/ha (47 gal/minute/ac) will be required. If 4 mm/hour ($\frac{1}{6}$ in./hour) is used 708 litre/minute/ha (63 gal/minute/ac) will be required. In calculating the water requirements of a plantation sufficient water should be available to provide for at least three and preferably more successive nights of frost, each of 10 hours duration, with provisions for a total of 60 to 80 hours operation during the season. A great saving of water can be made by taking advantage of any opportunity to recover the sprinkled water by running or pumping it back into a reservoir.

OPERATION OF A SPRINKLER SYSTEM

Good equipment and a very high standard of maintenance is necessary to avoid breakdown. The equipment should be set up and tested well in advance of the first expected frost. With so much at stake, the presence of an operator throughout the period of sprinkling is essential to ensure that the system is operating satisfactorily. Even with an automatic installation, pumps, motors, water pressures and sprinklers need to be checked. A powerful spot torch is useful to check each sprinkler at the commencement of sprinkling. Blocked sprinklers are best replaced by spares. This can be simplified by having stand-pipes incorporating automatic valves which seal when the sprinkler pipe is removed. If a pressure drop occurs during sprinkling and cannot be rectified it is advisable to cut off some lines rather than trying to continue to cover the whole area at the lower pressure. In maintenance, particular attention should be given to filters, which should be cleaned regularly.

ANCILLARY USES OF SOLID SET SPRINKLER SYSTEMS

In addition to frost protection and irrigation purposes (page 40), a sprinkler installation of this kind can be used for the application of nutrients by using an injector, or more simply by feeding the partly diluted material into the suction side of the pump. Pesticides have been applied in this way but control of diseases in particular has not been as good as that provided by normal spraying machines, which achieve much better leaf cover.

Soil

Black currants will thrive on a wide range of soils. The best results are obtained on deep, medium-heavy to light loamy soils. The main requirement is a deep soil that encourages free rooting, with a good supply of moisture during the summer months. A minimum depth of at least 0·5 m (18 in.) is required but 0·75 m (2 ft) is preferable and more if the soil is open textured. Although the greatest proportion of the black currant roots are in the top 0·5 m (18 in.) of soil many of the thicker roots will penetrate 1–1·3 m (3–4 ft) in good deep soils. Obviously these bushes exploit a greater bulk of soil than bushes with more restricted root systems and are better able to carry regular, heavy crops particularly under summer drought periods. On shallower soils irrigation may be essential and although this can be provided it adds to production costs.

Very close textured soils, provided they are not waterlogged in winter, are suitable for black currants grown under clean cultivation with herbicides. The ability to control weeds with herbicides, rather than frequent cultivations, now enables currants to be grown on very heavy loams and medium clays. But on very close textured clays growth is often slow, particularly when the bush is young.

Rather shallow soils over sandstone, ragstone and marl, provided the underlying strata are sufficiently porous, are reasonably satisfactory—shallow soils over chalk are not recommended; there may be sufficient summer moisture in the chalk but nutrient problems will arise associated with the highly alkaline subsoil. On the other hand the black currant is very intolerant of acid soils, more so than most other fruit crops.

The Agricultural Development and Advisory Service offers a chargeable service, as do some fertiliser firms, for the testing of soils. Preferably two samples each drawn from several cores over the field, from the top 15 cm (6 in.) and also from the 15–30 cm (6–12 in.) depths are used for the analyses. The pH gives the degree of acidity of the sample; pH 7·0 is neutral, below this the soil is acid and above this the soil is alkaline. Black currants in general will not grow satisfactorily below pH 6·0. Above pH 7·0 other problems occur.

Lime in various forms can be added to soil to correct low pH values, but on very acid soils, pH 5·0 or below, the large amounts of lime required may immobilise other plant nutrients. Thus very acid or very alkaline soils should be avoided.

Soils well supplied with organic matter generally have a good texture, drain well and yet have good available moisture capacities and a balanced supply of plant nutrients. Dressings of bulky manures, such as farmyard manure, broiler house and other litters can help to improve soils, but their use adds to establishment costs. Ploughed-out grassland and leys if carefully treated (page 30) before planting usefully add organic matter to the soil.

Proposed sites should be carefully inspected and the depth, texture and type of soil examined in several places overall using a soil auger. It is advisable to inspect fields in the winter when waterlogging or areas needing draining can be identified. It is possible to detect badly drained soils in the summer from colouration in the soil cores. This crop will not succeed on ill-drained soils and if natural drainage is unsatisfactory advice should be obtained as to whether field drainage is possible. If an existing tile drain system is in use ditches should be checked and outfalls inspected.

If a system of frost protection by overhead sprinklers is to be used (page 9) adequate field drainage is essential. Heavy soils are not suitable for frost protection using overhead sprinkler systems unless they are on sloping sites which will allow runoff, as if level, the plantation is liable to become water-logged after sprinkling.

Varieties

The main commercial variety is Baldwin and probably at least 70 per cent of the present area is down to this variety. Under suitable conditions Baldwin gives heavy yields, growth is compact compared with other varieties, the firm berries ripen evenly and hang well, are rich in vitamin C, acceptable for all types of processing and for the fresh fruit trade. Wellington XXX is the next most important variety, it ripens before Baldwin so is useful in lengthening the season, but the growth is rather lax and the berries sometimes split and ripen unevenly. It is widely grown in Kent. In some areas, particularly in East Anglia, French Black (syn. Seabrook's Black) is popular; it ripens a little before Baldwin. Westwick Choice is a very compact growing variety ripening after Baldwin and hanging well, but the yield is often disappointing.

There are several other varieties, some have been grown for many years but are not now important commercially. A few thousand bushes of each variety are certified each year, mainly for sale to amateur gardeners. These varieties are Amos Black, Blacksmith, Boskoop Giant, Cotswold Cross, Laxton's Giant, Malvern Cross, Raven and Tor Cross.

The earliest variety to ripen is Boskoop Giant; this is now not grown commercially as both Tor Cross and Mendip Cross ripen nearly as early and give heavier yields.

One variety is of particular interest. Hatton Black (syn. Green's Black) has been known for very many years but recently it has been found to carry a better crop than Baldwin on some frosty sites. Because of the possibility that the variety is less frost susceptible than Baldwin there is an interest in its propagation and testing.

Four other varieties are completely new and protected by Plant Variety Rights. One was raised by East Malling Research Station and named Jet. It is still being tested but is characterised by exceptionally long, pendulous strigs of fruit, which it was hoped would cheapen hand picking costs. Another new variety was raised at Long Ashton Research Station and named Blackdown. Both varieties are low in vitamin C and not suitable for processing. The other two varieties, raised at The Scottish Horticultural Research Institute, and named Ben Lomond and Ben Nevis are of great commercial interest since ADAS trials have shown that they outyield Baldwin. Ben Lomond is acceptable to the major processors and appears to have some inherent frost resistance and is less prone to mildew than Baldwin. Tests show that it can be harvested mechanically. This variety is now being widely planted to see if it can replace some of the Baldwin area.

The harvesting machines have picked most of the commercial varieties grown and there do not appear to be varietal differences; any problems that have arisen have generally been associated with type of bush and degree of ripeness of the crop, rather than with any particular variety.

With the advent of picking machines time of ripening has become important as there is a limit to the area of Baldwin that can be picked by one machine in one season. The industry really requires a variety with all the good characteristics of Baldwin but ripening about a week earlier, and another similar variety ripening a week later. In addition to the area down to Baldwin growers may now try to grow an area of the earlier ripening Wellington XXX and avoid uneven ripening by careful pruning to prevent the bushes becoming overcrowded, and by this means lengthen the harvest season.

For hand picking for the wholesale market Baldwin again forms the bulk of the crop, but a small area of Tor Cross and Wellington XXX may also be grown for early fruit when prices are higher. For late fruit Westwick Choice, Amos Black or Jet may lengthen the season when prices may again be higher. Raven is noted for very large fruit and the new variety Jet should be tried, as both look attractive in market containers.

Black currants for a pick-your-own enterprise will need to be associated with other, more popular, soft fruits. The early varieties will clash with strawberries and raspberries, so again the main area should be Baldwin with only limited quantities of early varieties. It is unwise to grow a later maturing variety than Baldwin as most customers will probably have obtained their requirements by the time Baldwin is harvested.

Varieties in usual order of ripening

Early	Boskoop Giant
	Laxton's Giant
	Mendip Cross
	Tor Cross
Midseason	Hatton Black (Green's Black)
	French Black (Seabrook's Black)
	Raven
	Cotswold Cross
	Wellington XXX
	Blacksmith
	Malvern Cross
	Blackdown
	Ben Lomond
	Ben Nevis
Late	Baldwin
	Westwick Choice
Very Late	Amos Black
	Jet

VITAMIN C CONTENT

The vitamin C content of black currant juice is now of less importance than before as artificial supplies may be added and the currant juice is required mainly for colour and flavour. The fruit of some varieties contain less vitamin C than others, but no particular variety is outstanding. The varieties can be divided into three groups according to the vitamin C content:

Low	Medium	High
Cotswold Cross	Mendip Cross	Boskoop Giant
Amos Black	French Black	Laxton's Giant
Hatton Black	(Seabrook's Black)	Wellington XXX
(Green's Black)	Ben Lomond	Blacksmith
Blackdown		Raven
Jet		Baldwin
Ben Nevis		Westwick Choice

Descriptive notes on varieties

Baldwin

An old variety of unknown origin. *Bush:* generally of medium vigour or less, varying with soil and locality, fairly compact. Commences growth very early and flowers early. *Trusses:* usually one or two per node; medium in length, each bearing six to nine berries. *Berry:* medium size, colour and flavour good; skin tough, hangs and travels well. *Season:* late. The main commercial variety—very satisfactory for processing.

Amos Black

Origin: Goliath × Baldwin. Raised in 1927 by Mr H M Tydeman, East Malling Research Station, introduced in 1951. *Bush:* vigorous and compact with stiff shoots; commencement of growth and flowering late. *Trusses:* one or two per node; rather short, each bearing about seven or eight berries. *Berry:* medium large; skin thickish, fairly tough, good colour, travels well. *Season:* very late, ripens a week to ten days after Baldwin. It is very late in flowering compared with all other varieties and may on occasion avoid frost damage. Cropping has not been high.

Ben Lomond

Origin: raised in 1961 by Mr M Anderson of the Scottish Horticultural Research Institute, Dundee and introduced in 1975. Protected by Plant Variety Rights 1975. *Bush:* medium vigour, upright but spreading with crop, buds are early to break, flowering after Baldwin. *Trusses:* usually one or two per node, medium in length each with six to eight berries. *Berry:* large, thick tough skin, sweet with less acidity than most black currants. *Season:* earlier ripening than Baldwin. Heavy cropping variety suitable for processing. Has cropped well in recent frost years and may be a suitable replacement for Baldwin. Suitable for mechanical harvesting.

Ben Nevis

Origin: raised by Mr M Anderson in 1961 at the Scottish Horticultural Research Institute, Dundee and introduced in 1975. Protected by Plant Variety Rights

1975. *Bush:* medium vigour, upright inclined to spread when cropping, buds early to break flowering a few days after Baldwin. *Trusses:* two to three per node, medium in length with seven to eleven berries per truss. *Berry:* medium-large, skin rather thick and tough; sweet, flavour fair with less acidity than most black currants. *Season:* earlier than Baldwin by about three days. Heavy cropping variety but not as heavy as Ben Lomond. Experience so far suggests that it will be suitable for jam manufacture but not suitable for juice.

Blackdown

Origin: Baldwin × Brodtorp. Raised by Dr D Wilson of Long Ashton Research Station in 1960 and introduced in 1971. Protected by Plant Variety Rights 1971. *Bush:* large, very spreading and fairly dense. *Flowering:* early midseason. *Trusses:* usually more than one per node with four to five berries. *Berry:* large, firm, good sweet flavour, easy to pick. *Season:* second early, a few days before Baldwin.

This new variety has cropped more heavily than Baldwin in a trial at the National Fruit Trials in Kent on young widely spaced bushes. It is much less susceptible to mildew than Baldwin. Not suitable for processing. Of main interest to amateur gardeners and for self-pick enterprises. Makes too spreading a bush for commercial use.

Blacksmith

Origin: raised by Messrs Laxton Bros (Bedford) Ltd. and introduced by them in 1916. *Bush:* vigorous, making a large to very large bush, rather spreading in habit. *Trusses:* usually one or two per node; moderately long, each bearing seven to ten berries. *Berry:* medium to large; good flavour and colour. Skin rather thin and tender, travels well. *Season:* midseason. Very little grown, mainly of interest to amateur gardeners.

Boskoop Giant

Origin: raised by Mr. Hoogendyk, Holland, about 1885 and introduced into England in 1895. *Bush:* vigorous, large and slightly spreading with a few strong main branches. Commences growth about midseason and flowers about midseason. *Trusses:* usually only one, occasionally two per node; usually long, bearing eight to ten berries. *Berry:* large; juicy, moderately sweet, with a thin, rather tender skin. *Season:* early. Only a moderate cropper. It is the earliest variety to ripen but fruit does not hang well and should be picked before terminal berries are fully coloured. It is rather liable to running-off, particularly in a cold spring, and is susceptible to leaf spot in the west of England. Not planted commercially today as Tor Cross is preferred.

Cotswold Cross

Origin: Baldwin × Victoria. Raised in 1920 by Mr G T Spinks, Long Ashton Research Station, and introduced in 1932. *Bush:* moderately vigorous, medium to large, tall with a slight spreading habit. Flowers early. *Trusses:* usually two per node: moderately long thick strigs each bearing seven to nine berries. *Berry:* medium to rather small, uniform in size, sub-acid, good colour, skin thick, travels well. *Season:* midseason, a few days before Baldwin. A variety that is not grown extensively. Picking is not as easy as most varieties because the rather short strigs are hidden in the truss leaves.

French Black (Syn. Seabrook's Black)

Origin: Introduced by Messrs W Seabrook and Son Ltd, of Chelmsford, in 1913. Identical with French Black. *Bush:* vigorous, fairly compact and upright, much branched. Comes into leaf late, flowers late-midseason. *Trusses:* variable, from one to three per node; medium length, bearing six or seven berries. *Berry:* medium in size; acid; skin medium thickness and toughness, hangs reasonably well and travels well. *Season:* midseason. A variety that was widely grown on various types of soil but mainly in the east and south-east of England. A good grower but with a tendency to running-off, and this may account for the only moderate crops often obtained.

Hatton Black (Syn. Green's Black)

Origin: Boskoop Giant probably crossed with Carter's Champion. Raised by Mr H Jones, Market Drayton in 1912. Another variety received from Greens of Norfolk under the name of Green's Black has been found at the National Fruit Trials to be morphologically indistinguishable from this variety. *Bush:* moderately vigorous, fairly compact, of similar size to Baldwin, produces a moderate number of one-year old basal shoots. Flowers early midseason, reaching full flower a day earlier than Baldwin. *Trusses:* two to three short trusses per node. *Berry:* of medium size, skin tough. *Season:* second early, ripening with French Black and Wellington XXX. It has cropped well in some trials and better than Baldwin in frost years. Although the fruit is reasonably well exposed, the short trusses do not lend themselves to easy hand picking, but it should be suitable for machine harvesting.

Jet

Origin: derived from *Ribes fuscesens*. Raised by the plant breeding section East Malling Research Station and introduced in 1973. Protected by Plant Variety Rights 1973. *Bush:* moderately vigorous and spreading with much lateral branching, a habit which may not be entirely suitable for machine harvesting. *Flowering:* sufficiently late for the risk of spring frost damage to be reduced. *Trusses:* usually there is only one per node of remarkable length, averaging 11·5 cm (5 in.) with 10–20 berries. *Berry:* variable in size mainly rather small, firm. *Season:* very late. The long strig is expected to improve the rate of picking by hand as compared with other varieties. The fruits detach without tearing the skin and are thus dry in the basket. There may be a gap between the picking of other varieties and the start of picking Jet. It was bred for ease of hand picking, and very late flowering to escape spring frost damage.

Laxton's Giant

Origin: Raised and introduced in 1946 by Messrs Laxton Bros (Bedford) Ltd. *Bush:* large, vigorous, spreading, rather sparsely branched; commencement of growth and flowering midseason. *Trusses:* one or two per node; medium length, each bearing nine or ten berries. *Berry:* very large; skin thick but tender. *Season:* early, a few days after Boskoop Giant. Generally only a moderate cropper. The berry is one of the largest and the truss ripens fairly evenly for an early variety. Of main interest to the amateur gardener.

Malvern Cross

Origin: Baldwin × Victoria. Raised in 1920 by Mr G T Spinks, Long Ashton Research Station, and introduced in 1946. *Bush:* large, tall, vigorous, fairly compact; leafing and flowering early to early midseason. *Trusses:* one or two per node; medium length each bearing six to eight berries. *Berry:* medium large, sub-acid, good colour, skin rather thick and tough. *Season:* late, ripening about the same time as Baldwin. The cropping of this variety has been very variable and some heavy crops have been recorded, but it has not proved reliable enough and excessive vegetative growth has also been a problem.

Mendip Cross

Origin: Baldwin × Boskoop Giant. Raised by Mr G T Spinks, Long Ashton Research Station, in 1920 and introduced about 1932. *Bush:* moderate in vigour on light soils, vigorous in areas with a high rainfall, cup-shaped, producing plenty of new wood, but lower branches droop to the ground and need removing by pruning. Early in leafing, flowering about midseason. *Trusses:* one or two per node; the primary truss bearing eight or nine berries. *Berry:* large, slightly acid, skin thin but fairly tough, ripens unevenly, does not hang well. *Season:* early, usually a few days later than Boskoop Giant. This variety has cropped well, particularly in the south-west, though it can be disappointing as it does not thrive equally well in all areas. It is difficult to handle, especially in a wet season, as the period available for picking is short and the fruit quickly becomes over-ripe. Tor Cross, although a few days later to ripen, is now considered a better early variety.

Raven

Origin: Boskoop Giant × Baldwin. Raised in 1905 by Messrs Laxton Bros (Bedford) Ltd and introduced by them in 1925. *Bush:* large to very large, with a few strong branches; bud break and flowering early midseason. *Trusses:* usually single; long, bearing 10–12 medium-large berries. *Berry:* large to medium, skin thin and tender. *Season:* early midseason. Has cropped well in recent experiments. One of the varieties more easily picked but does not hang well. Excellently flavoured juice.

Tor Cross

Origin: Baldwin × unknown variety resembling Boskoop Giant. Raised in 1924 by Mr G T Spinks, Long Ashton Research Station, and introduced in 1962. *Bush:* moderate in vigour, cup-shaped, producing an ample supply of new wood from the base. Flowers mid- to late season. *Trusses:* frequently in pairs; each bearing seven to eleven berries. *Berry:* medium to large, skin tougher than other early varieties and fruit hangs and travels better. *Season:* early, ripening just after Mendip Cross. A new variety yielding better than Mendip Cross in trials but which may be less frost susceptible and more even in ripening.

Wellington XXX

Origin: Boskoop Giant × Baldwin. Raised by Mr R Wellington, East Malling Research Station, in 1913 and introduced in 1927. *Bush:* vigorous, with a pronounced spreading habit; one of the earliest to produce leaves and early in flowering. *Trusses:* usually one or two per node: medium to long; rather crowded on the branch, the primary truss bearing eight to ten berries. *Berry:* medium to large, but small at the end of the strig, sub-acid, with moderately tough skin. Does not hang well and inclined to split and the berries can ripen unevenly on the strig and throughout the bush. *Season:* midseason. Crops satisfactorily under widely differing conditions and it can produce very heavy crops in favourable seasons. The chief disadvantages are frost susceptibility and the spreading habit of the bush. Has a tendency to uneven ripening particularly when bushes are too dense.

Westwick Choice

Origin: Raised by Mr G D Davison, Westwick Fruit Farm, Norfolk, and introduced by Col. Petre of the same address. *Bush:* moderately vigorous and fairly compact in habit; commencement of growth and flowering early midseason. *Trusses:* usually two to three per node; medium length. *Berry:* medium to large; sub-acid, skin rather thick and tough; fruit hangs well but not as long as Baldwin. *Season:* late, coming in just after Baldwin. Fruit has a tendency to split especially after rain. Grown mainly in East Anglia.

Pollination

All black currant varieties grown in Britain are self-fertile, that is they do not need pollen from another variety for fertilisation and seed, and hence berry, production. However the pollen grains from the anthers must reach the stigmas in each flower to effect fertilisation. Experiments have shown that bushes caged to prevent the presence of insects will only give a very poor crop. This indicates that, although some pollination can be ensured by wind, insects are necessary for good pollen movement.

Unfortunately the black currant flowers very early in the year during April and often during times when the weather is very changeable. Hive bees are seldom seen on black currant flowers. This may be because the air temperatures are frequently below 15°C (59°F) when this crop is flowering and hive bees are not very active below this temperature. Bumble bees are often seen foraging amongst black currants and are valuable as they shake the whole strig: other wild bees and insects may also be important in effecting fertilisation and the greatest care must be taken not to use insecticidal sprays that will reduce their numbers. Wild bees nest in warm, dry, grassy banks and hedge bottoms and may be encouraged by care in retention of suitable sites.

Black currant bushes produce far more flowers than necessary for a full crop. A recent survey in the West Midlands on the variety Baldwin calculated that a typical plantation of established bushes could produce some 89 million flowers per planted hectare (36 million/ac). If the flowers had set and produced typical sized berries, theoretical yields of 50 tonne/ha (20 ton/ac) would have been produced, instead of the normal crops. This super abundance of flowers ensures that, if some flowers do not get pollinated, or are damaged by frost, wind, insects or disease, sufficient flowers remain to produce the crop. It is thus natural to see some dead and dried up flowers, which soon drop off, on the strigs after flowering. The term 'running-off' is used to describe this character only when insufficient flowers have set to give a satisfactory crop.

Propagation

Commercially black currants are propagated by hardwood cuttings made from one-year old shoots planted in the ground during autumn and spring which root the following summer. The ease with which the black currant roots can often be seen in fruiting plantations where a shoot, at or just below ground level and covered with a little soil or mulch, will readily form roots during the summer, while still attached to the parent bush.

Because of the ease of propagation great care must be taken to select bushes that are true to name and free from diseases and pests. There is a particularly important virus disease colloquially called reversion that causes the bush to become entirely fruitless although the vegetative growth continues strongly (page 60). It is absolutely essential not to propagate from reversion-infected bushes, since the cuttings will carry the virus in the cell sap and the new bush will be as useless as the parent.

CERTIFICATION SCHEMES

It is mainly to help prevent the propagation of reversion-infected bushes as well as to ensure trueness to variety that the Ministry of Agriculture, Fisheries and Food many years ago started a voluntary system for the inspection and certification of most black currant varieties.*

There are two schemes. The Special Stock Scheme requires stringent isolation of the nursery area from all other currant or gooseberry bushes. It is limited to

---

*Information and application forms for the black currant certification schemes can be obtained on application to Plant Health Branch, Ministry of Agriculture, Fisheries and Food, 90–96 Cannon Street, London EC4N 6HT.

certain varieties, the stocks of which have to be provided through the Nuclear Stock Association from specially tested mother plants originating from the East Malling and Long Ashton Research Stations. The stocks are inspected twice during the growing season and there are limits to the number of years that the stock remains eligible for entry to the scheme. An exceedingly high standard of apparent freedom, at inspection, from pests and diseases and complete trueness to variety is demanded. No fruiting bushes are eligible; only bushes cut back (stooled) for the production of cuttings and two-year old bushes in nursery rows, that must also have been certified in their first year.

At present there are only one or two growers producing the main commercial varieties to SS standards. Certified Stocks are listed and the Register of Stocks is available from the Ministry each autumn.

Most black currant propagating material is provided through the A Scheme. Bushes of eligible varieties can be entered for certification if grown as:

1. Two-year or older bushes in nursery rows or
2. Bushes stooled for the production of cuttings or
3. Bushes in fruiting plantations that have not been planted for more than three years.

One-year old bushes cannot be entered for certification as, owing to their growth, it is impossible adequately to inspect them for trueness to type or freedom from reversion.

The stock entered must not be adjacent to any other plantation substantially infected with reversion virus disease or gall mite (page 60). Horticultural advisory officers of the Ministry inspect these plants entered for A certificate once during the growing season to ensure that the stocks meet the minimum requirements of the scheme. Stocks meeting these requirements receive a numbered certificate and a register of growers of these stocks is prepared each year and can be obtained from the Ministry, or consulted at divisional offices.

It is strongly recommended that only certified bushes or cuttings from certified bushes are used to establish plantations. A copy of the certificate should always be obtained from the supplier when purchasing stock.

SOURCE OF CUTTINGS

The best source of cuttings is from stooled bushes that are inspected and certified each year. Excellent cuttings can also be obtained from certified two-year old bushes when these are first planted and cut down to the ground. Cuttings may also be obtained from one-year old bushes when these are cut down for growing on as two-year old bushes in nursery rows, provided the one-year old bushes were raised from certified material.

Cuttings should not be taken from fruiting bushes as there is a great risk of pest and disease infection. However cuttings from fruiting bushes not more than three-years old after planting should be satisfactory if the bushes have been certified. Cuttings should never be taken from old plantations, or from non-certified bushes, and never from bushes showing the slightest signs of gall mite infection (page 55).

Stools are generally developed by planting certified two-year old bushes and each winter cutting off all the shoots made in the previous summer. This encourages the formation of new vigorous growths of one-year wood suitable for cuttings. All shoots should be removed from the parent bush each winter. It is important not to keep the stool bushes for too many years, or to manure heavily, otherwise they will tend to produce rank growth difficult to spray satisfactorily. Regular inspections should be made for black currant gall mite and reversion disease and the stools should be given a full spray programme to control pests and diseases. As no fruit will be produced on stool bushes, or one or two-year old bushes cut down, crop residues are not a problem and more effective spray materials against black currant gall mite may be used than is possible on fruiting bushes.

Provided the one-year old shoots are well lignified the whole length can be used to make cuttings. There is evidence that cuttings root best if inserted in autumn when the soil is warmer, than during the dormant winter season, or in spring. There is also evidence that cuttings made from any age of wood will root readily if planted in August and September, even if they are inserted with all leaves on, provided the soil is moist or irrigation can be provided. Cuttings so inserted may make a little growth in the autumn and excellent growth the following spring. Care must be taken to avoid the cuttings drying out before insertion. However the most usual time to take cuttings is during the dormant season. The cut off shoots must be bundled and if not used immediately should be well buried in moist soil in a cool place to avoid shrivelling.

The length of cutting depends on the method of insertion. If a planting machine is to be used the cuttings may have to be cut to about 18 cm (7 in.) long, as the machine can only insert to this depth to leave one bud above the soil. If the cuttings are to be hand planted in a plough furrow, or in a subsoil tine furrow, or using a spade, a longer cutting may be used up to about 30 cm (1 ft).

INSERTING THE CUTTINGS

Whilst many types of soil are suitable for the propagation of black currant cuttings the lighter and more sandy types will give the best results. The land should be well prepared and cultivated to a depth of at least 26 cm (10 in.) and be in a well broken condition throughout, but not so fine that the surface will pan or consolidate excessively with treading or winter rainfall; the site must not be ill-drained. Any deficiency in soil nutrients should be corrected by pre-preparation application of the appropriate fertilisers which are subsequently well cultivated into the top depth of soil. Soils well supplied with humus will aid root production and produce good bushes under drought conditions.

Although it is usual to state that the cutting should be prepared by trimming to just below a bud at a node, at the end of the cutting to be inserted in the soil, black currants root so readily this is often not carefully done under commercial conditions. The important point is to plant the cutting as deeply as possible, to trim back the top to leave only one bud above soil level and to ensure that the cutting is very firmly embedded in the soil. This is particularly important if cuttings are being pushed in a tine furrow by hand. On a small scale a spade can be used to open a narrow slit into which the cuttings are pushed deeply by hand.

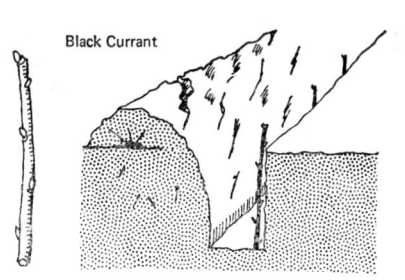 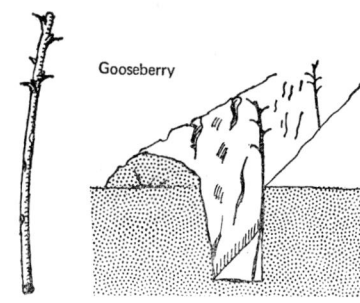

In the spring after planting, particularly if hard frosts have occurred, the cutting bed should be inspected to see if the frosts have loosened the cuttings. If so they should be firmed back, by foot on a small scale, or by using a light roller for large areas.

Cuttings are usually spaced 15 cm (6 in.) apart in rows 1 m (3 ft) apart in the nursery (and must not be closer than this if to be entered for the A Certification Scheme). This will permit of two year's growth in that site without overcrowding. The rows may be closer than this at 60 cm (2 ft) if it is intended to lift and plant out after only one year's growth in the nursery position.

TREATMENT AFTER PLANTING

Provided the soil has been thoroughly prepared the only weeds expected should be annual, and they can be controlled by the use of the residual herbicide simazine. This is applied either after planting, or in the early spring, at 0·8–1·1 kg a.i./ha* (¾ to 1 lb a.i./ac) and will kill most annual weeds as they germinate and absorb the herbicide. If simazine resistant annual weeds, particularly knotgrass, are anticipated lenacil may be used instead (page 45).

As the cuttings grow the new leaves and shoots will need protection by spraying against pests and diseases. In the first year there will not be much growth until mid-June, when routine sprays against gall mite (page 55) and mildew (page 59) should be applied. Leaf spot is seldom troublesome on young bushes.

Once rooting has taken place irrigation may be applied later in the season if the soil is very dry and shoot growth unsatisfactory. However it is important not to over-irrigate cutting beds as this may produce a soft unbalanced bush with lush shoots and poor roots.

At the end of the first year's growth the rooted cutting may either be lifted to be planted out, or more usually, be left to grow on another year. In this case the new shoots are all cut off in winter to leave only one bud per shoot situated above soil level. This is very important as it encourages a well stooled bush with plenty of shoots arising at or below soil level and avoids a 'leggy' plant with a length of old bare wood between soil and the lower shoots. The black currant readily produces shoots from buds which grow out below soil level, on the original cutting wood, or rootstock (Plate III).

---

*a.i./ha—active ingredient per hectare.

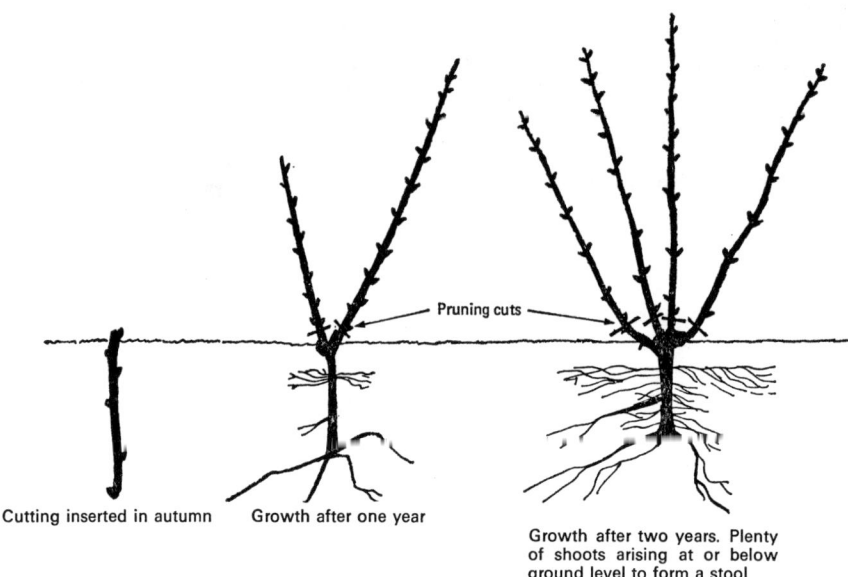

Cutting inserted in autumn   Growth after one year

Growth after two years. Plenty of shoots arising at or below ground level to form a stool

In the second year the nursery bush should make much more growth, producing perhaps three to eight good strong shoots. An adequate programme to protect these shoots against gall mite, mildew and if necessary aphids must be carried out.

In the second year all the nursery bushes must be carefully inspected to ensure that they are of the correct variety and not showing symptoms of infection with reversion virus disease. Any doubtful bushes should be dug out and burnt.

Rooted one or two-year old bushes are usually dug out by hand on a small scale. On a larger scale it is usual to plough out the rows of bushes, or to lift them with an undercutting nurseryman's tine. The bushes, once lifted, should not be left lying in the field as the roots will dry out. Surplus soil is shaken off and weak bushes graded out to be destroyed, or planted back for one more year's growth in the nursery row. Depending on size the bushes are then bundled and tied up in constant numbers and then should either be covered with sacks and consigned in lorry loads, or the roots covered in soil in rows to await collection and planting.

Lifting and replanting is best done as soon as the leaves have fallen in the autumn. This is not always possible and bushes can be lifted and planted from October until March so long as soil conditions permit and the bushes remain dormant.

IDENTIFICATION OF VARIETIES AND REVERSION VIRUS DISEASE

Cuttings and plants are liable to become mixed in commercial handling and a constant watch is therefore needed to keep the stock true to type. Growers with sufficient experience will be able to rogue for trueness of variety at the same time as they rogue for reversion virus disease (page 60).

Correct identification of black currant varieties can be accomplished only after considerable experience in the field. A number of distinctive characters help in identification; for example, during the dormant season the buds of French Black and Boskoop Giant are red while those of Baldwin are brownish-pink. Suspected mixtures may therefore be sorted out during the winter by reference to this character. Varietal identification is best done during the summer, from mid-June to late July, at the same time as roguing for reversion, using a key to stem and leaf characters. The features of particular value are the colour of the stem and leaf stalk, pubescence (very fine hairiness) of the stems, and the depth of the basal sinus (the indentation of the leaf blade where the stalk joins it). To give an example: nursery rows of Boskoop Giant and Wellington XXX may appear very similar, but these varieties can be distinguished by the colour of their stems and leaf stalks, Boskoop Giant having green stems and purplish-red leaf stalks, while both stems and leaf stalks of Wellington XXX are reddish-brown to purplish-red. Similarly, Mendip Cross may be confused with Wellington XXX in the nursery row, but it can be distinguished by the dense hairiness of the stems, compared with the slight hairiness of Wellington XXX and the difference in colour of the stems. Full details of the morphological characteristics of many black currant varieties are given in Reference Book 411 *Black Currant Varieties* (see Appendix I).

Experience is also needed to detect reversion-infected nursery bushes by their leaf characters (Plate VII). This identification can be done at the same time, late June to late July, that the bushes are inspected for trueness to variety. It is also recommended that new fruiting plantations, including those established from certified stocks, should be inspected the first and second year after planting to confirm that all bushes are of the correct variety and disease free. Any doubtful bushes should be removed and burnt. Thereafter it will probably be more easy to identify any reversion-infected bushes in the spring at the time of flowering rather than on leaf symptoms in the summer (Plate I).

Planning the plantation

Black currants are grown in rows. The width between the rows is determined by the vigour expected from the bushes and the type of tractor and mechanical harvester to be used. On most soils an alley width of 2·7 m (9 ft) is the minimum that will permit the passage of tractor and sprayer, or harvester, without damage to the adjacent bushes. If the soil is good, or irrigation is to be used, the alleyway may have to be a little wider at about 3 m (10 ft).

Many experiments and the experience of growers have shown that if bushes, or cuttings, are planted closely in the rows the yields in the first two or three years are higher than from more widely spaced plantations, as there is a greater amount of fruit bearing wood per unit of ground. In particular the modern aim is to ensure continuous rows of bearing shoots, rather than distinct bushes. This is necessary for mechanical harvesting and avoids the production of shoots growing up or down within each row that prove difficult to pick. It is also satisfactory for hand picking.

Baldwin, the main commercial variety, is a fairly compact grower, though it can make a more vigorous bush on good soils well supplied with water, or if irrigated. Bush spacings that suit Baldwin will also be satisfactory for French Black, Westwick Choice and Tor Cross. However Wellington XXX will make a taller bush more inclined to droop fruit-laden branches into the alleyways and this variety may need slightly wider row spacing than Baldwin under the same conditions. Cotswold Cross and Raven will also need extra spacing compared with Baldwin.

BUSH AND ROW SPACINGS

Given freedom to grow in all directions a black currant bush will form a rounded bush; the area covered and the height increasing with age. Old bushes, if not kept pruned, become very leggy and the fruit is only carried on the young wood at the tips of the branches which will droop both across the alleyways and into adjacent bushes. Thus a large bush is not required and bush to bush spacing combined with pruning should be planned to give continuous rows of fruiting shoots, suitable for both hand and machine picking.

Trials under very good growing conditions at ADAS Experimental Horticulture Stations have shown the benefits in the first and second crops from close in-row spacing. By the third year onwards there was little difference in yield from bushes planted differentially but the total accumulated crop was highest from bushes originally planted as cuttings 30 cm (1 ft) apart.

Before the advent of herbicides black currant bushes were spaced in the rows to allow for hand hoeing between bushes for weed control. Many plantations are still planted with bushes 0·9–1·2 m (3–4 ft) apart but this is too far apart if bushes are to be mechanically harvested, although cheaper in bush costs at the time of planting.

Irrespective of the planting system adopted, either cuttings, one or two-year old bushes can be planted. Because of their cost and bulk two-year old bushes are generally not used to establish three row bed systems, or for the close in-row spacings, for which one-year old bushes, or cuttings are more suitable.

It is possible to obtain certified two-year old bushes and these should always be planted in preference to uncertified bushes. Experienced growers, planning their planting programme some years ahead, are able to place a definite order for bushes which helps the propagator and may enable a discount to be given over prices ruling at date of dispatch. Certified one-year old bushes are not available as explained (page 21) but if one-year old bushes are ordered it should be stipulated they must have been propagated from cuttings from certified stock, or from certified two-year old bushes. In all cases the supplier should be asked to quote the appropriate certificate number on the invoice for identification purposes.

Where two-year old bushes are planted strong growth should result in the first two years and this type of bush may be planted not more than 0·9 m (3 ft) apart in the row, or closer if plenty of bushes are available; 0·75 m ($2\frac{1}{2}$ ft) is often adopted.

If one-year old (yearling) bushes are used a plantation can be more quickly established if planted about 0·6 m (2 ft) apart in the rows. Naturally the more bushes planted per unit area, the higher the establishment costs. If a plantation is to be established by planting unrooted cuttings these should be closer.

MULTIPLE ROW OR BED SYSTEMS

Many growers aim to get higher yields in the first few years by planting bushes on a bed system. The most usual method is to plant three rows 1·5 m (5 ft) apart from each other with a 3 m (10 ft) alleyway for the tractor passage. The centre row of bushes is reached by spraying through the outside rows. Herbicides are applied over the top of the bushes from an offset sprayer arm and fertilisers are spun over all the bushes. The bushes are spaced within the three rows at normal distances. The crop is hand picked for one, two, or three years until the bushes are high enough to be mechanically harvested, or become too crowded. In the last year the central row is picked by hand and then this row of bushes can be quickly pulverised, not rotohoed, allowing either the mechanical harvester to pass over the remaining single rows left at the 3 m (10 ft) spacings, or the hand pickers to get through more easily.

The three row bed has proved successful where the static headland picking machine is used, as the bushes can be kept severely pruned to provide fruiting branches for the machine (page 65). With hand picking three row beds can be retained for as long as the bushes are not overcrowded and the pickers can work amongst the bushes. This might be for four to six crops. The centre row should always be removed before inter-row competition becomes apparent on the bushes.

Four and five row beds have been tried but are not recommended as they generally become too thick and it has been found difficult to obtain adequate control of pests and diseases on the inner rows. It is also more difficult to thin these systems to leave well spaced single rows, than with the three row bed.

Number of bushes required

Per hectare				Per acre			
Distance in-row (m)	Between rows (m)			Distance in-row (ft)	Between rows (ft)		
	2·7	3·0	3×1·5 m bed		9	10	3×5 ft bed
0·45	8230	7410	11 120	1½	3230	2900	4360
0·60	6180	5560	8340	2	2420	2180	3270
0·75	4940	4450	6680	2½	1940	1750	2620
0·90	4120	3710	5570	3	1610	1450	2180
1·00	3710	3340	5010	3½	1390	1250	1870
1·15	3220	2900	4350	4	1210	1090	1640

PLANTATIONS DIRECT FROM CUTTINGS

The method of establishing plantations direct from cuttings rooted *in situ* is strongly recommended only where experience shows a high percentage take of cuttings and good growth can be expected. This method is *not* recommended for growers inexperienced in the propagation of black currants, or on fields or soils, where because of lack of water, irrigation or weed problems, propagation would be difficult, expensive or unsatisfactory.

The method enables excellent growth to be made as the rooted cuttings are not subjected to a transplanting check. Where irrigation is available, or water reserves of the soil are good, it may be possible to crop the cuttings in the first year, that is the second year after planting the cutting, and not to cut them down to the ground, as is necessary when transplanting one or two-year old bushes.

The only disadvantage of rooting cuttings *in situ* is that, if the cuttings have to be cut down, the area of ground occupied by the cuttings in the first year is greater than if the cuttings had been raised in a nursery bed, and subsequently transplanted. But the transplanting costs and check to growth are avoided.

An absolute essential is to ensure the use only of healthy cuttings. Because of the high number of bushes produced any unhealthy cuttings will be more difficult to detect and remove. Cuttings purchased from producers of certified stools, or cuttings from the shoots of certified two-year old bushes cut off after planting should be used. Cuttings obtained when one-year old bushes are planted and cut down are satisfactory, provided these one-year old bushes have been raised from certified cuttings under known reliable conditions.

Preparing, inserting and propagating cuttings is described on pages 21–24. Exactly similar methods are used when the cuttings are to be inserted in their fruiting rows. Generally the cuttings will be inserted by using a planting machine or by ploughing-in. As the rows will be permanent it is important that they are lined out very accurately in the field and that the soil has been deeply cultivated and is free from all perennial weeds. Good preplanting preparation of the soil and correction of any nutrient deficiencies must be undertaken before planting the cuttings (page 30).

For the best results the cuttings should be planted in the row, fairly thickly. Cuttings set out at 90 cm (3 ft) apart, for example, are too wide. Not all cuttings will root, there may be a 10–20 per cent failure and at these stations the gaps will be very wide and need replanting the following autumn. It will also take several years before continuous rows are obtained.

Provided plenty of cuttings are available, for best results plant not wider than 30 cm (1 ft) apart in rows 3 m (10 ft) apart, or 30 cm (1 ft) apart in three row beds as described above. If insufficient cuttings are available but percentage rooting and resultant growth are expected to be really satisfactory, cuttings may be inserted at 45–60 cm ($1\frac{1}{2}$–2 ft) apart.

If insufficient cuttings are available it would be wiser to plant at 30 cm (1 ft) distance over half the area and plant the remainder of the area the following year, from cuttings taken from the first lot. The second area could be planted a little closer in the row to use up all the available cuttings. This would give a well grown level plantation in the first cropping year.

Number of cuttings required

Per hectare				Per acre			
Distance in-row (cm)	Between rows (m)			Distance in-row (in.)	Between rows (ft)		
	2·7	3·0	3×1·5 m bed		9	10	3×5 ft bed
15	24 700	22 230	33 330	6	9680	8710	13 070
22	16 840	15 150	22 730	9	6460	5810	8710
30	12 350	11 110	16 670	12	4840	4360	6530
38	9750	8770	13 160	15	3880	3480	5230
45	8230	7410	11 110	18	3230	2900	4360

ROW LENGTHS

Irrespective of the type of material used to plant the field it is essential to consider the continuous length of each row. If the crop is to be picked by hand very long rows over 130 m (150 yd) will mean that pickers will have to walk at least half this distance to reach a headland for weighing the fruit. Long continuous rows however help reduce tractor operation time. But harvesting is the biggest cost and most likely bottleneck and so if very long rows fit best down a field, it is wise to leave cross alleyways every 90 m (100 yd) or so, to permit the passage of a tractor and trailer for fruit removal.

Planning the length and arrangement of rows if machine harvesting is to be used is even more important. All the mobile machines need wide headlands of a minimum of 6·7 m (22 ft) in order to turn smoothly, and the row length should be adjusted to the type of harvester to be used (page 65). The length of the row should be such that when the plantation is in bearing in a normal year the machine passing down one complete row will harvest not more than the maximum sledge or trailer capacity of the machine. If more than this is harvested the machine will have to stop to off-load full trailers in mid-row, or work out partial rows.

It is difficult to give information on suggested row lengths for machine harvesting. The best method is to assume an expected average crop on the chosen in-row and alley spacing and to calculate the row length that will yield the maximum capacity of the particular mobile machine to be used. Even so in light crop years the trailers will not be completely full, whilst in heavy crop years it will be necessary to stop the harvester down the row to off-load full trays and pick up empties. With this calculation to hand it may be possible to plan long rows with appropriate cross alleys so that in earlier or light crop years the machine passes down the whole row before the trailer capacity is full, while in heavy crop years the capacity is approximately reached when the machine is adjacent to the cross alleyway or alleyways. Long continuous rows if carefully planned in this way should give maximum efficiency for work with mobile harvesters and reduce both headland turning time and headland wasted area to a minimum.

Row length is not so critical with the static harvesting machine, where fruiting wood is cut and carried to the headland situated machine. But long rows are undesirable and rows about 70–90 m (75–100 yd) long, similar to those for hand picking, should prove suitable.

In addition to considering row length relative to the harvesting method to be used, the slope of the field and problems of air drainage for frost or irrigation pipe lay-out need to be considered, also protection from exposure and possible soil erosion problems on the steeper slopes. A plantation is only planted once, mistakes cannot thereafter be rectified, so the greatest care should be taken to select the best possible layout.

Planting

SOIL PREPARATION BEFORE PLANTING

The land must be physically and nutritionally suitable before planting and every effort must be made to control perennial weeds (page 48) as correction after planting is difficult, expensive and may reduce the weight of early crops.

Good growth can be expected if planting after a ley provided it is thoroughly disced or rotohoed before ploughing in. Perennial weeds should be controlled by herbicide treatment in the ley before it is broken up.

Soils used for arable crops are perfectly satisfactory for black currants provided texture and structure have been preserved. The aim should be to plan a suitable rotation such that the black currants are planted after a cereal crop, allowing adequate pre-planting soil preparation, rather than after sugar beet, potatoes or late winter vegetables which may leave the soil in a poor physical condition. In these cases it would be preferable to allow for a summer fallow before planting.

Old permanent grassland will need careful attention both for the control of perennial weeds and to ensure that the turf is not ploughed in as a mat but first thoroughly broken up.

Subsoiling is recommended on all sites to break down any cultivation or soil pans. It should be done under the driest possible conditions in order to ensure maximum fissuring of the soil. (See Leaflet 617 *Subsoiling*, Appendix I.)

If soil samples reveal that plant nutrients or lime are required (page 35), it is a good plan to apply half the quantities recommended before ploughing, to plough in and apply the remainder on the surface, thus ensuring good distribution through the top depth of soil. This is particularly important if lime is needed.

Except where a ley or grassland is ploughed out it is an advantage to apply farmyard manure, or broiler house litter, or any other bulky organic manures if available before planting. This type of manure is particularly valuable on lighter soil where it encourages the development of a strong root system. Also because it improves texture and moisture holding capacity.

After ploughing the land should be left in the furrow until just before planting when it should be cultivated, but then only enough to leave a rough surface once planting has been completed. Over-cultivation will lead to poor surface conditions particularly if heavy rains occur. Winter rain and frosts usually break down the soil surface after autumn planting to leave the surface suitable for an early spring herbicide application.

However if the land is very uneven after ploughing adequate cultivations or even cross-ploughing may be necessary to give suitable levels. On most soils it should not be necessary to cross disc or roll fields intended for bush fruit, especially as these operations may excessively compress the lower layers of soil. The aim in pre-planting cultivation is to ensure a friable 15–20 cm (6–8 in.) soil depth and avoid any panning below this. A chisel plough cultivator is a good tool to achieve well cultivated soil of good depth without too fine a surface. A seed tilth is not required.

Experience has shown that a plantation may be established following cereal crops without ploughing. This is only recommended if the site is free from perennial weeds, no nutrient deficiencies are present and adequate soil preparation—including subsoiling if required—was done prior to the cereal crop. It has the advantage that the cereal roots help maintain a stable soil structure and the stubble helps in surface stability, so that the actual planting takes place on a good soil surface. Annual weeds may be readily controlled by a suitable contact herbicide in the autumn prior to planting—the stubble should not be cultivated—and planting can then take place as soon as bushes can be lifted in the autumn. Cuttings can also be planted in a stubble in a cultivator tine slit. The uncultivated soil is marked out, and planted exactly as if cultivated.

MARKING OUT

Once the soil has been cultivated the plantation will need marking out. As herbicides are now always used for weed control there is no need to space the bushes very accurately in the row, but it is essential to get the rows at the correct distance to each other and exactly parallel.

On a field scale the most usual method is to line out a row of poles across the field, either on one side, or down the centre in an uneven shaped field. A reliable tractor driver should then be able to plough a straight furrow, or mark with a cultivator tine at measured distances away from this line using one or more sighting poles at each headland, to mark the row positions.

On a small scale a measuring tape and line can be used to mark the rows.

To mark the position of the planting holes for the bushes in the row it is hardly necessary to measure very accurately. On a field scale a piece of wood cut to the correct length will enable the planters to space out the holes. Planting two-year old bushes the stick should be 5–7 cm (2–3 in.) shorter than the actual distance required, to allow for the thickness of the bush, or the workers must be reminded to place the stick in the centre of the bush to measure onwards. Similarly the measurement should be made to the centre of the hole if a post hole auger is used.

If cuttings or yearling bushes are being planted fairly closely these are often set out by eye, or they may be measured with a stick as just described.

PLANTING

Planting is possible at any time provided the bush or cutting is dormant, generally from October to March. Late spring planting should be avoided; early autumn planting is best. The black currant starts growth very early in the spring and late planting causes a check to this growth. The soil must be friable and planting must be delayed if the soil is wet, sticky or frozen.

On a small scale the bushes can be planted by hand using a spade or fork to dig out the hole and throw the soil back around the roots where it should be firmed by the foot.

On a field scale the bushes may be placed in a ploughed furrow and the soil put back by use of a spade or fork. The disadvantage is that the soil between bushes also has to be replaced. If bushes are fairly widely spaced it may be possible to do this mechanically by cross cultivation, but this requires that the bushes are carefully sited in the rows. Alternatively the plough can be used to turn the furrow back against the bushes along the whole row length and a final soil covering by hand, together with firming of each bush, completes the job. This method is very suitable for cuttings or one-year bushes planted fairly closely in the row.

Similarly a potato ridger body can be used to open a furrow deep enough to plant one-year bushes or cuttings, but it may not be deep enough for strong two-year old bushes, where some hand work may be necessary to deepen the hole.

A post hole auger 30 cm (12 in.) is an extremely useful tool to prepare planting holes. It must never by used under wet conditions, particularly on heavy, or clay soils as it then smears the side of the hole, forming an impervious surface through which the bush roots cannot grow. Care must also be taken to adjust the auger to give the right depth of hole. If there is any doubt about the surface of the auger hole the workers should break down the sides a little with a fork before placing the bush in the hole and shovelling back the soil. A post hole digger is excellent for planting two-year bushes, but of no use for one-year old bushes if planted more closely, or for cuttings.

When planting great care should be taken not to allow the bush roots to become dry. On delivery from the nursery the bushes should be heeled in, placed in a trench or furrow and soil thrown over their roots to keep them moist, if they cannot immediately be planted. Bushes are usually tied up in bundles as delivered from the propagator. Damp sacks should be placed over the bush roots on the trailer to keep them moist before they are dropped out in their planting positions. Do not drop bushes too far ahead of the planters.

If bushes have been propagated on the same farm it is well worthwhile lifting only sufficient bushes for planting each day. This avoids keeping the bushes out of the ground for longer than is necessary.

The bush is placed in the hole at the correct depth and friable soil thrown back around the roots. Two-year bushes with extensive roots should be quickly jogged up and down to ensure that the soil falls well down around the lower roots. The soil should be well firmed with the foot, further soil added and firmed to leave a slight mound above the surface; this will allow for some natural shrinkage of soil around the bush and avoid leaving it in a depression.

The depth of planting is very important. The bushes should be planted deeply, slightly lower than originally in the nursery rows to ensure good growth of shoots from below soil level. Never plant a black currant bush to leave a single leg with shoots arising above this. If a nursery bush has such a leg plant it deeply to bury the base of the one-year shoots.

After planting some soils may need a little between row cultivation—under dry conditions—to level out tractor wheel marks and leave the soil surface suitable for the life of the plantation without any more cultivations.

Planting machines can be used for cuttings and, provided the soil is very workable and the roots are not too large, for one-year old bushes. Two-year old bushes are too large to be planted with a machine. It is important to ensure that the bushes or cuttings are planted deeply enough and that the press wheels firm the soil well around each bush.

When planting cuttings to remain in their fruiting positions in the field they can be placed against a ploughed furrow wall, or in the furrow made by a ridger and the soil thrown back by hand and the cuttings firmed by pressure of the two feet, one on each side. Cuttings can be planted in a deep tine mark by being forced downwards in the slit to press on the bottom and then being firmed. They can be firmed back by running a tractor wheel along the row to press the soil.

On a small scale cuttings are inserted by hand using a spade to make a slit about 20 cm (8 in.) deep. The cutting is pushed in to reach the bottom of the slit and be in contact with the soil—this is important—and the slit closed back upon each cutting by firm pressure of the foot. After planting all cuttings should be cut off to leave only one bud above soil level. Some growers prefer to insert cuttings at a slight angle in the soil in the belief that this gives a longer length of stem in the soil and so improves rooting.

All cuttings may need re-firming in the spring if frost occurs during the winter after planting. This is seldom necessary for well rooted bushes as the weight of soil around the roots should prevent any frost heaving. However if severe frosts have occurred after planting it would be wise to check that the bushes are firmly in contact with the soil, in the early spring, and refirm if necessary.

Manuring

As the black currant flowers and fruits on shoots produced the previous year adequate supplies of moisture and plant nutrients are essential to maintain good growth. At one time the black currant was considered to be a gross feeder and to require plentiful supplies of bulky organic manures as well as fertilisers. But these requirements were quantified under systems of soil cultivation and often when bushes were competing for soil moisture and plant nutrients with weeds, both under the bushes and in the alleyways. Cultivations by tractor and hand to control weeds also damaged many of the bushes' fibrous roots that are mostly in the top 45 cm (18 in.) of soil, and so reduced soil moisture and hence nutrient uptake.

Nearly all black currants are now grown in undisturbed soil, all weeds being controlled by herbicides. Root damage is no longer inflicted and weed competition should be negligible. Under these conditions there is now considerable evidence to show that the nutrient requirements of black currants can be more accurately assessed and satisfied.

MAJOR PLANT NUTRIENTS

The major plant nutrients required for satisfactory growth are nitrogen (N), phosphorus (P), potassium (K) and magnesium (Mg). Lime or calcium (Ca) is also required as is a range of minor elements, particularly iron (Fe) and manganese (Mn). These minor elements are generally present in adequate amounts in all soils and are seldom in short supply. It is the major elements, together with water, that have to be maintained to ensure continuing good growth and cropping.

SOIL MOISTURE

It is important to realise that all plant nutrients have to be in solution to be taken up by the roots of the bushes. Soils will vary greatly in their capacity to provide moisture, throughout the spring, summer and autumn, to the bush. Medium loamy soils of good structure and texture well supplied with organic matter will provide the highest possible supply of soil moisture. This available water capacity (AWC) can be measured and is often expressed in terms of the moisture available in the different horizons, or depths, of the soil. The coarser the soil, the more sandy it feels, and the less organic matter it contains, the less soil moisture will be available. Thus the physical state of the soil, as well as the expected summer rainfall should be borne in mind when considering the manurial requirements of this crop.

ORGANIC MATTER (*Leaflets 435, 320)

This term is used loosely to describe a whole series of products which range from undecayed plant and animal tissues through the whole range of their breakdown processes to reach the state known as soil humus. The process is proceeding all the time. The soil is a living medium, a delicate balance of the living microfauna and microflora whose activities release plant nutrients into forms suitable for uptake by the plant roots.

Growers' experience, if not experimental data, suggest that soils liberally supplied with organic matter in the form of farmyard manure, broiler or other litters, compost, suitable sewage sludge, or plant debris from ploughed leys or grassland produce better growth of black currants than soils not so well supplied.

Organic matter will also help in both conserving and releasing soil moisture, and on some soils where plant nutrients are not limiting, the value of bulky organic manures is probably related to the improvement in the soil moisture supplies rather than in the direct supply of the plant nutrients.

Provided organic matter can be thoroughly incorporated into the top 30 cm (1 ft) of soil before planting any form of it can only do good. But bulky organic manures are both difficult and expensive to handle and it is doubtful if their expense can be justified in economic terms on soils already in good physical condition.

---

*Advisory Leaflet—see Appendix 1, page 104.

Any form of bulky manure is suitable provided it is well rotted. Very fresh broiler or farmyard manure should not be used, but left stacked to decompose further when it can be applied as a surface mulch in the following years.

Straw or sawdust to be used as a surface mulch may absorb too much nitrogen from the top of the soil, to the detriment of the bushes. To avoid this an addition of 10 kg (22 lb) nitrogen per tonne (ton) of straw, or sawdust, should be added to the mulch to aid in its decomposition *in situ*. For the same reason these unrotted materials should not be ploughed in before planting.

If sewage sludge is available as a pre-planting soil dressing to be ploughed in to increase the organic matter, care must be taken to ensure it is of a suitable quality and does not contain heavy metals, or other residues, that may be toxic to plant growth.

Depending on availability 50–250 t/ha (20–100 ton/ac) of bulky manures may be ploughed in before planting. A tonne (ton) of well made farmyard manure (FYM) contains approximately 1·5 kg nitrogen, 2 kg P_2O_5, 4 kg K_2O and 0·8 kg Mg. All composts and bulky manures vary greatly in their analyses, depending on the original materials and methods of decomposition.

AMOUNTS OF PLANT NUTRIENTS

Plant nutrients in fertilisers are expressed in percentages in the weight of fertiliser purchased. Thus a 50 kg bag of compound fertiliser with an analysis of 10 N : 5 P_2O_5 : 7 K_2O would contain 5 kg N, 2·5 kg P_2O_5 and 3·5 kg K_2O. The elements NPK are always given in the same order. By law in the United Kingdom all fertilisers, but not bulky organic manures, have to show their analyses on the container or invoice note.

SOIL SAMPLING (Leaflet 270)

It is possible to determine the levels of phosphate (P_2O_5), potash (K_2O), magnesium (Mg) and lime in the soil through soil analysis. As the black currant crop occupies the ground for some ten to twelve years it is strongly recommended that a careful soil analysis is made prior to pre-planting soil preparation, so that remedial action can be taken as necessary.

An auger is used to draw soil cores at random over the field from two appropriate depths, the cores are mixed and the two representative samples are then analysed in the laboratory. ADAS offers a chargeable service both to draw the samples and for their analyses. If the field is large, over 2 ha (5 ac), or if previous cropping has been dissimilar, soil samples should be taken from known areas so that, if necessary, differential fertilisers can be provided to parts of the field.

The soil is analysed in a standard manner and the results are reported as indices for the main elements, but not nitrogen which varies so much depending on season that no results are meaningful for this element. If required the organic matter can also be determined and results are expressed as a percentage. The indices range from 0 (very low) to 10 (exceptionally high). In soils of good fertility indices of 2 to 4 are normal. Organic levels generally range from 1·5 to 3·0 per cent.

If the soil has been analysed prior to planting it is recommended that it should be reanalysed every three years during the life of the plantation. The soil cores should be drawn from positions in the alleyways and below and between the bushes to give a representative sample. Soil sampling can be of particular value in monitoring soil pH levels, also for phosphorus and potassium. Intelligent use of soil analyses combined with leaf analyses may enable more economical annual fertiliser programmes to be used than those normally recommended. It will also be possible to check whether liming within the plantation is necessary during the life of the bushes. Such a requirement may arise on soils that quickly acidify.

LIME (Leaflet 518)

The black currant is very intolerant of acid soil conditions and a careful soil analysis is essential to determine the nature of the soil. This is expressed as a soil pH. A pH of 6·5 indicates a slightly acid soil, below 6·5 indicates an acid soil and above 7·0 an alkaline soil. At below pH 6·0 black currants will generally not grow satisfactorily. On the other hand soils with pH 6·8 or above may give rise to problems, as some of the trace elements such as iron and manganese may become less available. A soil with a pH 6·5 is ideal. If the soil sampling shows levels below this figure the necessary amount of lime to apply to bring the soil to pH 6·5 will be specified. If the analysis also shows low, or rather, low levels of magnesium the opportunity should be taken to apply the correct amount of calcium in the form of magnesian limestone, if this is available.

If the lower depths of soil are more acid than the top depth the lime should be ploughed in. If the soil is fairly acid in both depths apply half the lime required and plough in and the remainder on the surface to be cultivated in.

NITROGEN (Leaflet 441)

The amount of available nitrogen varies from day to day in the soil, being derived as a by-product of the activities of the soil micro-organisms and so cannot be assessed by soil analysis. The availability of nitrogen depends markedly on soil moisture and particularly summer rainfall (April-September). Thus the ADAS recommended nitrogen rates vary according to the average rainfall expected in the locality. If the summer rainfall is below 350 mm (14 in.) apply 140 kg/ha (110 units/ac) N per annum. If summer rainfall is over 350 mm (14 in.), or if irrigation is used, apply 70 kg/ha (55 units/ac) N per annum.

These recommended rates can be reduced if bulky manures are used. There is increasing growers' experience to suggest that once the bushes have grown to fill the space in the rows these rates may be unnecessarily high. Leaf ash analysis (page 38) can then be used as an additional guide to maintenance of fertiliser application.

PHOSPHORUS (Leaflet 442)

Phosphorus is needed particularly for root production and it is therefore essential to ensure adequate, even generous, supplies at planting time. Phosphorus can be immobilised in certain soils, particularly acid ones, so maintenance dressings may be required annually. However, when small maintenance dressings only are required (index 2 or 3), twice the annual rate can be applied biennially.

Soil sample nutrient index	P_2O_5 annual applications	
	kg/ha	units/ac
0	110	90
1	70	55
2	40	30
3	40	30
Over 3	Nil	Nil

Phosphorus is normally supplied in the form of the straight fertiliser superphosphate or triple superphosphate, or in granular compound fertilisers together with nitrogen and potash. It is important to use fertilisers containing soluble forms of phosphorus; insoluble forms are too slow acting for bush fruits.

POTASSIUM (Leaflet 443)

This element is particularly important for most fruit crops and is often deficient on the more sandy soils. Annual maintenance applications are required on deficient soils. At index 2 or 3 twice the annual rate can be applied biennially.

Soil sample nutrient index	K_2O annual applications	
	kg/ha	units/ac
0	250	200
1	180	145
2	120	100
3	60	50
Over 3	Nil	Nil

Potassium may be supplied in the fertilisers sulphate of potash, or muriate of potash. In most compound fertilisers the potassium is in the form of the muriate. This is not detrimental to black currants, although it is not recommended for red currants or gooseberries when large amounts are required on deficient soils when the sulphate form should be used.

MAGNESIUM (Leaflet 444)

The black currant has a moderate requirement—in comparison with some other crops—for magnesium but if the pre-planting soil analysis gives an index of 2 or below magnesium may be required. The method of application will depend on whether lime is also required, in which case magnesian limestone (3–12 per cent Mg) may be used. If lime is not required magnesium may be supplied as Epsom salts (10 per cent Mg) or Kieserite (16 per cent Mg).

Soil sample nutrient index	Mg annual applications	
	kg/ha	units/ac
0	60	50
1	40	30
2	30	25
Over 2	Nil	Nil

There is a relationship between potassium and magnesium in the soil and their respective uptake by plants; if potassium levels are high this can reduce the uptake of magnesium by the bush. These two elements therefore need to be assessed and, if necessary, adjusted together.

MINOR ELEMENTS

Deficiencies of iron (lime-induced chlorosis), manganese or other minor elements should not occur if black currants are grown on fertile soils with a pH of about 6·5.

Any plant nutrient can be in short supply if the bush roots cannot properly exploit a satisfactory block of soil. Thus on shallow or waterlogged soils, or in soils over impervious rock pans, bushes may show leaf and growth deficiency symptoms. These are due, not to an absolute deficiency of the plant nutrient or nutrients, in the soil, but to the poor rooting, or to lack of oxygen caused by waterlogging. Similar symptoms can be caused by fungi attacking the root system.

Deficiencies of minor nutrients seem to be rare, though black currants suffer occasionally from lime-induced chlorosis in soils with a high pH, or sometimes in parts of fields where lime-containing water seeps through, or again where there may be outcrops of lime-containing rocks such as cornstone, or on soils over chalk. Little can be done except to avoid liming, to use acid fertilisers and, where the drainage is at fault, to correct this. Deficiencies of other minor elements are generally associated with the overall problem of growing black currants on very poor or waterlogged soils and their correction seldom renders the plantation profitable.

LEAF ASH ANALYSIS

Mature leaves from the centre of the current year's shoots should be collected just before the fruit is ripe. A representative sample from bushes all over the plantation is needed. These leaves must then be dried within a day and the leaf digests are used for laboratory determination of the major plant nutrients nitrogen, phosphorus, potassium, magnesium and calcium. Thus leaf ash analysis is an accurate way of determining the plant nutrients present in the bush, particularly at the time of flower initiation (July) when yield potential for the following year is determined.

Considerable experience is now available, based on experimental and advisory work, to enable the results of leaf ash analysis to be used both as a guide to manuring and as a help to the solving of nutrient problems. The ADAS offers a chargeable service both to collect the samples, which needs to be carefully done, and their analysis and interpretation.

The information available from leaf ash analysis is of particular value when the plantation is four or five-years old, as it can be a guide to the fertiliser programme for the next period. There is ample evidence to suggest that established black currants, if planted on soil of reasonable depth with adequate soil pre-planting preparation, may not need the annual rates of fertiliser suggested in the preceding paragraphs. In such cases the savings on unnecessary fertiliser applications should cover the leaf analysis service.

It is suggested that black currants with a leaf ash analysis in July of nitrogen 2·8 to 3·0 per cent, phosphorus 0·25 to 0·35 per cent, potassium 1·5 per cent and magnesium 0·15 per cent have ideal nutrient levels for growth and crop, providing water supply and climatic conditions are satisfactory. Leaf ash analysis is not particularly useful for calcium, which is best monitored by soil analysis.

Where leaf ash analyses show results above or below these levels fertiliser applications of the particular nutrients may be adjusted accordingly. It is important to relate analyses to the weather conditions particularly for rainfall, or to irrigation if used. Under different soil moisture levels uptake of plant nutrients could alter.

Leaf ash analysis also assists in keeping the right balance between the different elements, for example it would show if potash manuring has been so high that the uptake of magnesium has been limited.

If as a result of leaf ash analyses fertiliser applications are stopped, adjusted, or reduced, further leaf ash analyses should be done in two years time to monitor progress and ensure that soil nutrients are still available in adequate amounts for bush uptake.

Regular routine winter soil sampling about every three years, combined with leaf ash analyses as required, should enable nutrient levels to be adjusted to the optimum levels for growth and crop.

Irrigation

When the supply of water for the bush is inadequate it can reduce the crop which is to be picked in the year of drought and also the potential crop for the next year, because of the reduction in the shoot growth made during the drought. In most areas few seasons go by without some drought period, which may or may not come at a critical time, and some addition to the natural water supply is usually an advantage. The few experiments that have been carried out on the irrigation of black currants generally show the advantage of adequate supplies of water at the right time and emphasise the long-term effects of drought. During dry seasons crop increases of 50 per cent from both pruned and unpruned bushes were achieved at East Malling Research Station by applying 4·5 cm (2 in.) of water when the available soil moisture reserves had been reduced by 4·5–6·0 cm (2–2½ in.), even on the Station's relatively deep soil of a fine sandy loam texture.

Where water supplies for irrigation are limited, smaller applications, or only a single application applied early in June, should be effective. Obviously the response to irrigation will depend very much on the soil and the average summer rainfall, particularly during June and July.

Good growth from newly planted, or young, bushes and particularly the growth of cuttings, can be severely checked under abnormally dry spells. Irrigation may then be of particular value not so much to improve a small early crop, but to ensure the good shoot growth essential for the next year and succeeding crops.

Fruit size and firmness are generally improved by irrigation without reducing the vitamin C content. Some delay in fruit ripening may occur on irrigated bushes, particularly on lightly pruned bushes, though this should be less than a few days. More of a problem is that vigorous, irrigated bushes may give uneven ripening of the fruit on the strig; experience is needed in irrigation applications to avoid this. There is also more possibility of *Botrytis* rot infection of the berries of irrigated bushes, particularly if they are allowed to become over-ripe.

It is difficult accurately to assess the effects of irrigation on the occurrence of diseases. There is some possibility that irrigation may increase the amount of leaf spot through dispersal of spores (page 58) but may reduce the incidence of powdery mildew (page 59).

Irrigation equipment is expensive and if its necessity can be avoided by choosing sites and situations where, except in exceptional seasons, soil moisture is adequate, this will reduce costs. However where soil moisture is the only limiting factor, or where the frost risk justifies the installation of sprinkler equipment, then irrigation may be economically justified.

SPRINKLER EQUIPMENT

The apparatus described on page 9 for sprinkling for frost protection can also be used for irrigation. The pipes and standpipes are already in the rows, so water can be applied as and when required. In many cases the only limiting factor will be the quantities of water available.

All water supplies are now subject to control and growers considering the use of irrigation equipment for bush fruits should, in the first instance, consult their local Drainage and Water Supplies Officer of the Ministry of Agriculture at the divisional office. He will be able to explain the necessary procedure to obtain a licence to use water for irrigation purposes, the particular problems of supplies in the locality and advise on construction of reservoirs and other matters.

Full details of the different types of irrigation equipment available, including pumps and the construction of reservoirs and use of equipment are given in Reference Book 138 *Irrigation;* also Reference Book 416 *Potential Transpiration* (Appendix I) describes methods of calculating the soil moisture deficit for any area, provided the local rainfall is accurately known.

Many growers use mobile pipes and sprinklers which can be moved from crop to crop. The movement of even lightweight aluminium alloy pipes and sprinklers in and out of bush fruit plantations is not so easy as for overground crops. Thus the solid set frost protection sprinkler irrigation systems once installed are much easier to operate, but the capital cost is higher compared with mobile systems.

Irrigation through the solid set system can be more frequent and so allow a more careful relationship between irrigation, the rainfall and the total deficit. With mobile systems growers generally apply at least 3 cm (1 in.) of irrigation, often 6 cm ($2\frac{1}{2}$ in.) at less frequent intervals. To apply 3 cm/ha requires 300 000 litre (1 in./ac requires 22 000 gal). Using these figures the actual water requirements per area per season for different amounts of water to be applied can be calculated. Under farm conditions it may not always be possible to apply irrigation as frequently as is desirable. Alternatively, if heavy applications are made soil erosion may become a problem on sloping sites.

Care should be taken to fit the irrigation applications to the spray programme. Heavy irrigation will wash off most protective fungicides from the leaves, but after irrigation it may be a few days before the tractor and sprayer can travel between the bush rows without damage to the wet soil. Similarly mobile pipes cannot be immediately moved from watered plantations without damage to the wet soil.

Irrigation applications should be accurately determined by flow meters or by timing and the information recorded. This will enable local experience to be factual and help in future years.

Irrigation may be needed after picking if conditions have been very dry, to encourage and continue the growth of the new shoots, but should not be continued after early September, to allow for cessation of shoot growth and ripening of the terminal shoot tips and buds.

Pruning

The fruit of the black currant is borne mainly on the young shoots produced during the previous growing season, the aim of pruning fruiting bushes is to remove the older growth and retain the younger wood and also to encourage the annual production of sufficient vigorous young growths. Pruning may be done at any time during the dormant season but before the buds start to swell in the spring. There is no disadvantage in pruning early in the autumn before leaf fall, or any time after picking.

PRUNING TOOLS

Secateurs are used for the first two years of pruning, thereafter it is easier and quicker to use hand loppers, or secateurs with 46 cm (18 in.) handles. Depending on age and size of bushes it takes about 50–70 man hours/ha (20–30 man hours/ac) to hand prune black currants. Bushes kept properly pruned from their early years are easier to handle than those not pruned each year, or where old branches have been left in. Removal of thick old branches is not easy with hand loppers and will slow the work.

Pneumatic secateurs, power-assisted by air lines from a compressor on the tractor pto, have often been used in conjunction with the static headland harvester. It is usual to have a gantry over one row on each side of the tractor enabling one complete and one half row to be pruned on each side of the tractor; generally one compressor will power four secateurs. These machines are noisy to work with but can increase pruning speed and are worth considering if large areas of bush fruit have to be pruned, irrespective of the method of harvesting used.

Some growers have endeavoured to use various types of farm hedge trimmers, generally the reciprocating knife type, automatically to prune up the sides of established bushes. The knife arms are adjusted at an angle to trim the shoots close to the bush at the base. Unfortunately this method of pruning tends to remove the potential fruit bearing wood and leave the ageing and unthrifty wood untouched in the centre of the bushes. It is only of value on old or neglected plantations where the bushes have become intermingled across the alleyways and should be followed by selective hand pruning in the centre of the bushes.

PRUNING AFTER PLANTING

After planting, the bushes should be cut down to ground level. If the rows need to be marked for any operation on the land before the spring growth shows their position, a few sticks (shoots) can be stuck into the soil to indicate the rows.

All the shoots must be cut off to leave only one bud per shoot above soil level. This is very important. It prevents the bush cropping in the year after planting and induces growth of plenty of strong shoots at, or below, soil level, thus forming the basis of the stool (Plate III).

Much experience has shown that it is impossible to crop the black currant and also make satisfactory growth in the year after planting. However on a small scale, or in a garden, provided the bushes can be watered, it may be possible to thin out the shoots of strong two-year bushes, planted under good conditions, to leave a few shoots to crop the first year after planting, without hindering the growth of the new shoots. This is not suggested for commercial conditions.

AT THE END OF ONE YEAR'S GROWTH

At the end of the first season strong one and two-year old bushes should have made at least three to four and sometimes up to ten or more good long shoots. If growth is strong all that is required is to cut out weak or short shoots, cutting to one bud above ground level, leaving the other shoots to bear the crop, but often this is not possible under large scale conditions. If less than four to five shoots have grown, or if the shoots are weak, all the shoots will have to be cut down to ground level again. Obviously this is unsatisfactory, for if sufficient growth cannot be achieved in this first year cutting down again will delay the crop for a year. But strong growth is needed or the first crop will be too small to be worth picking and unpicked fruit weakens the successive growth. If cuttings have been used to establish the fruiting plantation, no pruning will be necessary unless growth is weak, in which case all the cuttings must be cut down to the ground.

AFTER THE FIRST CROP

Provided growth has continued satisfactorily plenty of new shoots should have been produced. The bush now consists of several two-year old shoots that have carried the fruit and are thicker and darker in colour than the light brown one-year shoots. On a field scale it is unusual to prune bushes of this age unless the growth has been unsatisfactory. In this case some of the two-year old shoots should be cut off, leaving one or two buds only at ground level. This will encourage more growth but reduce the fruiting wood for the next crop.

AFTER THE SECOND CROP

If cuttings or bushes have been planted closely, as recommended in this reference book, by this time the growth of the bushes should have resulted in rows of shoots with no spaces between. Pruning will be as described in the following paragraph for established bushes. Bushes planted at wider spacings will not yet have met in the rows, so the pruning must still be very light, only removing low shoots and those crossing or rubbing in the centre of each bush to encourage overall growth.

PRUNING ESTABLISHED BUSHES

Once the bush rows have filled the aim in pruning is to maintain plenty of growth of new shoots and remove the older branches before they have become too thick and heavy with much bare wood at the base and with the fruit only carried at the top.

First remove any broken or damaged branches and then some of the oldest branches, cutting these at or near ground level. If two to five such branches are removed each year a suitable rotation of cutting and regrowth will have been initiated within each bush—or length of row. After this major branch removal it may be necessary to make cuts away from the centre of the row, to 'set up' any drooping branches by pruning back to a suitably placed vertical shoot. Very tall or straggling branches may need shortening back to suitable vertical one-year old shoots. Finally crossing or crowded shoots will then need thinning out. The pruners should be instructed to make some cuts at or near soil level, as, if all cuts are made higher up in the bush few shoots will arise from the stool and the centre of the bush will not be rejuvenated. This pruning out of older branches must remove some of the young shoots arising from these old shoots but this cannot be avoided.

Varieties vary in their growth and some varieties produce far fewer and stronger branches with less shoots than Baldwin, others produce more, but unbranched, shoots. Mendip Cross and Raven and to a lesser extent French Black and Tor Cross produce fewer shoots than Baldwin*. Westwick Choice makes a very short compact bush whilst Cotswold Cross and Wellington XXX will need removal of lower shoots and the 'setting up' and stiffening of other shoots. Jet makes a very wide straggling bush and all spreading branches will have to be pruned away, or the fruit will touch the ground.

Bushes may be left unpruned in the last one or two years before grubbing as the maintenance of continuing new shoot growth is not then so important.

PRUNING BUSHES FOR MECHANICAL HARVESTING

Bushes to be mechanically harvested by mobile harvesters should be pruned exactly as if to be hand picked. The only difference is that strong shoots that lean from bush to bush in the row, rather than slightly out in the alleyways, are wrongly placed for most types of harvester and will need removal.

Bushes that have been closely planted in the rows for mechanical harvesting should present no problems. The aim in pruning is to maintain tall fruiting wood, as the machines cannot harvest fruit within about 25 cm (10 in.) of the ground. The machines gather in drooping branches very successfully, so bushes need not be 'set up' quite so much as when pruning for hand picking. Branches that have been broken, or that have been badly bark damaged by the passage of the machine in the previous year, need pruning out.

---

* Ben Lomond is similar in growth to Baldwin but spreads more with heavy crops.

DISPOSAL OF PRUNINGS

In the first few years the light one or two-year old shoots pruned out from the young bushes may be left in the alleyways—if not required for propagation.

When considerable quantities of older branches and shoots are being pruned out they are too much to be left on the ground and will get in the way of other operations. The prunings may be pushed out of the rows by using a narrow buck rake or rear loader on a tractor. If this is to be done the pruning gang may throw the larger prunings into every other alleyway. Prunings can then be pushed into a heap and burnt.

Alternatively a rear mounted pto pruning pulveriser can be used to pulverise the prunings *in situ* and leave them on the ground. Various machines are available but off-set types cannot be used in the bush rows and many grass flailers are not strong enough to pulverise the thicker branches so purpose-made pulverisers are necessary. Plant pathologists do not consider that thoroughly pulverised prunings form a disease source; the pulverised material gradually rots down and thus returns to the soil.

GRUBBING

Grubbing of old plantations can be an expensive operation if each bush is pulled out separately by a tractor or winch. Generally a tractor-driven rotavator is used and has reduced the cost of this task considerably. The rotavator is driven down the rows of bushes breaking both branches and roots. Usually this needs repeating three or four times, lowering the level of the rotary blades each time. The site is then left in a level state with the bush and roots broken into short pieces. This method is suitable before corn or other crops which do not need much row crop cultivation. The well broken pieces of bushes will decompose completely in a few years.

Control of weeds

Traditionally weeds were controlled by hoeing between the bushes and with cultivation between the rows. Placing a mulch of straw or other suitable organic material along the rows helped to prevent the growth of annual weeds but it was necessary to make sure that perennial weeds which would grow through the mulch were killed beforehand. Some growers applied a straw mulch overall on clean soil and thus avoided the need for any cultivation.

The use of chemical herbicides has now replaced cultivations in most commercial plantations. It is important that all perennial weeds are eliminated before planting for, although annual weeds are relatively easy to deal with, perennial weeds are more difficult and expensive to control in young plantations.

Trials have shown that there is an increase in yield from non-cultivated plantations compared with those that are cultivated. Although the soil surface appearance is less pleasing than with cultivated soil, the soil structure immediately below is often improved. After some years, the surface becomes firm and is frequently colonised by mosses. The nomenclature adopted to describe the weeds is that recommended in *English Names for Wild Flowers* as published by the Butterworth Group for the Botanical Society of the British Isles.

THE PRINCIPLES OF CHEMICAL WEED CONTROL

The herbicide used must kill or control the weed without damage to the crop. This selectivity depends on the plant tolerating the herbicide, or through avoiding contact between the herbicides and the plant. Tolerance is essential with soil-applied herbicides. Luckily bush fruits are tolerant of a wide range of herbicides and completely weed free plantations can be achieved. The occasional resistant weeds and seedlings of brambles and forest trees, for which no safe herbicides are available, must be dealt with by hand labour with the minimum of soil disturbance while they are small.

The herbicides used must provide adequate weed control within the known tolerance of the crop and yet the dose be low enough so that the chemical does not accumulate in the soil from year to year. With contact and growth regulator type herbicides there is little or no bush tolerance and safe use depends on directed sprays which avoid contact with the bush. As herbicides generally control only a limited range of weeds, a programme of several different herbicides is required over the years to avoid the build-up of weeds resistant to one particular herbicide within the plantation.

It is particularly important to use spot treatment on small patches of individual weeds within the plantation before they have time to become established. In particular established perennial weeds such as nettles, docks, thistle, couch grass, buttercup and bindweed can be readily controlled with specific herbicides which it would be both expensive and unnecessary to apply overall and which may put the crop at unnecessary risk. Some of these herbicides are not recommended for use until the bushes are well established, that is after at least one year's growth following the initial transplanting, or after cutting down.

THE CHEMICALS

All doses mentioned here are in terms of active ingredient (a.i.) per sprayed area. The standard chemical name is used. All the proprietary forms of these chemicals are listed in the annual booklet *Approved Products for Farmers and Growers* obtainable as a priced publication from HMSO, P.O. Box 569, London SE1 9NH, or through booksellers. Booklet 2264 (formerly HSG 25) *Chemical Weed Control in Bush and Cane Fruits* is revised more frequently than this Reference Book and should be consulted (Appendix I).

Since the 1976 edition of this publication there have been several changes in the herbicides Approved for use in bush fruits. Dalapon can be used for couch grass control. Lenacil is a safe residual herbicide sometimes used when weeds resistant to simazine are expected, and also on cuttings. A new Approved herbicide is pentanochlor, a contact material. Glyphosate is no longer Approved for use within bush fruits, but remains a valuable herbicide for clearing land of perennial and annual weeds prior to planting. Oxadiazon (not Approved) can be used for the control of bindweed, cleavers and some other simazine resistant annuals.

CONTACT HERBICIDES

Paraquat

This herbicide rapidly destroys all green tissue but is inactivated upon contact with the soil and has no residual activity. At doses of 0·6–1·1 kg a.i./ha (½–1 lb a.i./ac) it is effective in controlling annual grasses and broad-leaved weeds. Cleavers are somewhat tolerant and best results are obtained by spraying when two whorls of true leaves are present. The control of knotgrass and fumitory is also better when the first true leaves have formed than it is at the cotyledon stage. The addition of extra wetter will also improve the control of these weeds. As paraquat is not translocated within the plant to any great extent, it is not effective against deep rooted perennial weeds, with the exception of creeping buttercup. Paraquat is rapidly absorbed by plants; rain falling soon after spraying should not affect the results.

Paraquat may be used as a directed spray at any time of the year, but extreme care must be taken to avoid wetting the crop foliage or new shoots otherwise severe injury results. Gooseberry bushes can be sprayed overall with paraquat in the dormant season but this should be avoided with black currants as the buds remain green throughout the winter and may be injured by the spray.

RESIDUAL HERBICIDES

Simazine

At doses of 1·1–2·2 a.i. kg/ha (1–2 lb a.i./ac) this cheap and effective herbicide will give good control of a wide range of annual weeds, provided that it is applied before they germinate. Knotgrass, orache and cleavers tend to be resistant and the repeated use of simazine, by eliminating susceptible species, encourages a build-up of these weeds. Where this is starting to occur alternative herbicides must be used.

For simazine to be effective the soil must be moist. During dry times irrigation may be helpful to activate the herbicide, but this should generally not be necessary in bush fruits if the chemical is applied during February to March. The duration of weed control depends on the dose, type of soil and on ensuing weather conditions. An application of 1·1 kg a.i./ha (1 lb a.i./ac) of simazine may persist for up to six months. Increasing the dose will prolong the residual activity, but excessive doses are not necessary for control of annual weeds. In order to avoid damage to crops grown subsequently simazine should not be used in the year before grubbing a plantation. Black and red currant and gooseberry bushes and cuttings may be treated with simazine from the time of planting onwards. Simazine may be mixed with an overall paraquat application to kill annual weeds if already present in a newly planted and cut back black currant plantation.

Diuron

Diuron is a useful alternative material to simazine for use in black currants and gooseberries established for a year. At 2·4 kg a.i./ha (2¼ lb a.i./ac) it controls weeds such as knotgrass, fool's parsley and orache which are mainly resistant to simazine. It must not be used on light soils.

Chlorpropham

At doses of 2·2 kg a.i./ha (2 lb a.i./ac) chlorpropham gives effective control of autumn and winter germinating weeds, particularly knotgrass which is resistant to simazine. Chlorpropham will kill chickweed seedlings by contact action but for the control of other species it must be applied to clean land. Chlorpropham has a rather short residual life, especially during warm weather. Because of its limited persistence it may be used instead of simazine in the final year before grubbing a plantation. Groundsel and other composite weeds are resistant to chlorpropham but can be controlled by mixtures of chlorpropham and fenuron. The addition of fenuron widens the range of weeds controlled.

Chlorthiamid

Chlorthiamid is a residual herbicide which will control most germinating weeds as well as a range of established perennial weeds. It is formulated as $7\frac{1}{2}$ per cent granules and used at rates from 6·7–9·2 kg a.i./ha (6–$8\frac{1}{4}$ lb a.i./ac). The lower rate is sufficient for annual weeds, higher rates are required to control perennials such as thistles, docks, nettles, couch grass, coltsfoot and horsetail. Woody weeds such as brambles and black nightshade and annual cleavers are resistant. Creeping buttercup and bindweed are only suppressed and need separate treatment.

Chlorthiamid is recommended for use only on black currants and gooseberries established at least two years and should be applied during February to March. It can also be used in the early autumn against established perennial weeds, providing the soil is moist and the temperature is not too high. In hot dry conditions weed control is less effective. This herbicide, although volatile in warm weather, can be very persistent particularly if cultivated into the soil: it must not be used for pre-planting weed control, or applied to cut down bushes. It is very useful for spot treatment of patches of perennial weeds in plantations where simazine has been applied overall.

Dichlobenil

Dichlobenil is a residual herbicide closely related to chlorthiamid and must only be applied to established bushes. It is formulated as $7\frac{1}{2}$ per cent granules and is used at a rate of 6·5–11·2 kg a.i./ha ($5\frac{1}{4}$–10 lb a.i./ac). The lower rate will control annual weeds and the higher a range of perennial weeds with improved persistence. Application is normally in March or early April, but should be made before the leaves develop as the granules may lodge in the leaf and bud axils where they can cause damage. Dichlobenil is volatile and so should not be used near hops or greenhouses. Spot treatment of perennial weeds can usefully be achieved with dichlobenil. Do not apply to cut down bushes.

Propyzamide

Propyzamide is a residual herbicide effective as both pre- and post-emergence sprays against a wide range of annual weeds, notably annual grasses and chickweed. Cleavers and speedwells are also controlled but by higher doses. It does not control composites such as mayweed and groundsel. Couch grass, dock seedlings and creeping buttercup are also susceptible but bindweed, clover, perennial nettle and thistle are resistant. Propyzamide may be used at doses of between 0·8–1·7 kg a.i./ha ($\frac{3}{4}$–1$\frac{1}{2}$ lb a.i./ac) in bush fruits established for at least one year, the highest dose being required for couch grass, perennial weeds and difficult annuals such as cleavers. For successful weed control, particularly of couch grass, application should be made between 1 October–31 January. As the chemical is readily degraded under warm soil conditions, application earlier than October may leave insufficient effective residual life. When applied in January it is not likely to give effective control of susceptible annual weeds beyond May. For this reason, and that of obtaining control of a wider range of weeds, it is best used in a sequential programme with simazine applied in February.

TRANSLOCATED HERBICIDES

MCPB

MCPB in medium to high volume sprays can give a useful control of bindweed, thistle and creeping buttercup with little risk of injury to bush fruits if applied during August–September after extension growth has ceased, but before autumn frosts, at doses of 2·2–3·4 kg a.i./ha (2–3 lb a.i./ac). As far as is practicable the spray should be directed so as to avoid the crop, using only sufficient liquid to wet the weeds without causing excessive runoff to the ground and avoiding spray drift onto the bushes.

SUGGESTED HERBICIDE PROGRAMMES

All doses are given in terms of active ingredient (a.i.) per sprayed area
Pre-planting treatments

The programmes suggested below for established plantations will usually control most annual weeds but the control of perennial weeds in a crop is much more difficult. For this reason it is essential to achieve a good control of perennial weeds *before* planting. For the most common weeds such as couch grass, docks and thistles there is no one herbicide, cleared under the Pesticides Safety Precautions Scheme for use in bush fruits, which alone will give complete control. It is possible in theory to clean ground thoroughly of certain weeds such as couch grass, by means of a bare fallow which receives adequate cultivations and is favoured by dry weather, but this is not always achieved in practice and the best approach is to use a combination of chemicals and cultivations. On the fallow when cultivations are not possible weed growth may be checked by spraying with paraquat at up to 2·2 kg/ha (2 lb/ac). Where couch grass is the main weed a dose of 0·3 kg/ha ($\frac{1}{4}$ lb/ac) should be used. This is sufficient to kill existing shoot growth and stimulate dormant underground buds into growth. High doses may prevent regrowth for a considerable time, thereby reducing the efficiency of subsequent treatments, or cause them to be delayed. A low pressure sprayer will reduce the risk of drift, but where this is not important higher pressures will result in a better control of weeds.

Healthy (*left*) shoot of black currant at blossom time. Reverted (*right*) shoot showing hairless flowers and gall mite infected buds

Photo: *East Malling Research Station*

Photo: *East Malling Research Station*

Close-up of strigs of black currant showing difference in hairiness of healthy (*left*) and reverted (*right*) flowers

Well grown two-year nursery bushes being inspected for certification by an ADAS officer prior to sale and transplanting

The establishment of a black currant plantation from different types of planting material showing growth four years after planting.
Top: cuttings planted 0·6 m (2 ft) apart
Centre: one-year old bushes planted 0·6 m (2 ft) apart
Bottom: two-year old bushes planted 1·2 m (4 ft) apart

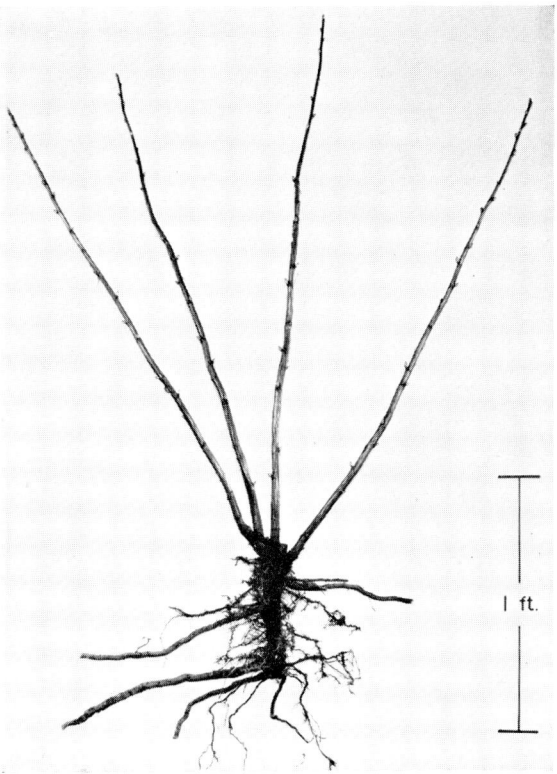

Two-year old black currant bush suitable for planting

Photos: *Long Ashton Research Station*

After planting cut off all shoots to leave only one bud. This will give strong growth in the second year from the stool

Copyright Bruff Engineering Co.

Tractor-mounted harvester

Copyright Smallford Planters Ltd.

Single sided mechanical harvester

Copyright Pattenden Engineering Co.

Pattenden self-propelled straddle harvester

Copyright Rubery Owen Group Services Ltd.

Smallford Hydrapick self-propelled straddle harvester

PLATE V

Before pruning

After pruning

Photos: *Long Ashton Research Station*

Blackcurrant bush, five years old, before and after pruning

PLATE VI

Black currant leaf spot

Photo: *L. G. Fish*

Photo: *East Malling Research Station*

Healthy leaf (*left*) and reverted leaf (*right*) of black currant. The main veins of the large central lobe of each leaf are numbered

PLATE VII

Modern gooseberry plantations in two row beds in herbicide treated soil. Note windbreaks

Gooseberries are often grown under plum trees in herbicide treated soil

Infestations of couch grass may also be killed by TCA-sodium, EPTC or dalapon. The first two are applied to the soil in the previous autumn or spring prior to planting. Their effectiveness depends on adequate soil incorporation to bring the chemical into contact with actively growing buds on the couch stolons. Dalapon, though having some activity through the soil, is mainly a leaf absorbed herbicide and so must be sprayed on to the actively growing foliage.

Where broad-leaved weeds as well as grasses are present both can be controlled with aminotriazole applied to actively growing foliage at a dose of up to 4·5 kg/ha (4 lb/ac). MCPA or 2,4-D may be added to enhance the control of broad-leaved weeds at a dose of 2·2 kg/ha (2 lb/ac). In the spring a year before planting a good tilth should be prepared and be sprayed overall with 2·2 kg/ha (2 lb/ac) simazine in order to control annual weeds. The perennial weeds can then be allowed to grow undisturbed so the correct herbicides for perennial weed control can be chosen and suitable sprays applied overall, or as spot treatments. By the autumn a high degree of weed control should have been obtained. If the soil is then in a suitable condition, further ploughing and cultivation may not be necessary and the plantation can be established on the weed-free, undisturbed ground.

Post-planting treatments

The following programmes should be read in conjunction with the information on the chemicals given on the previous pages. The programmes are not rigid and should be varied to suit local conditions and the presence of specific weeds.

After planting cuttings and bushes. Simazine at 0·8–1·1 kg/ha ($\frac{3}{4}$–1 lb/ac) in early spring after planting but before weeds germinate. This treatment is safe on medium and heavy soils. If simazine resistant weeds, particularly knotgrass, are likely to be troublesome, lenacil at 1·8 kg/ha ($1\frac{1}{2}$ lb/ac) could be used instead of simazine at the growers' own risk, followed by simazine applied in summer to early autumn if fresh weed seedlings emerge as the lenacil disappears.

Established fruiting plantations

February to March: If annual weeds have already germinated paraquat as a carefully directed spray at 1·1 kg/ha (1 lb/ac) for the control of these weeds may be necessary.

February to April: To keep the soil weed free during the remainder of the year simazine should be applied at 1·1–2·2 kg/ha (1–2 lb/ac) as an overall spray or diuron at 2·2 kg/ha (2 lb/ac).

November to December: Chlorpropham at 2·2 kg/ha (2 lb/ac) or chlorpropham plus fenuron 0·6 kg/ha ($\frac{1}{2}$ lb/ac) may be required but only if a heavy flush of winter germinating knotgrass is anticipated. From October 1 to January 31 inclusive propyzamide can be applied at 1·7–3·4 kg/ha ($1\frac{1}{2}$–3 lb/ac). The lower dose for the control of knotgrass, the higher rate for couch grass and cleavers.

The above programme may be repeated annually but usually simazine and paraquat resistant weeds gradually become dominant and perennial weeds may start to appear. In anticipation of this chlorthiamid or dichlobenil at 6·5–9·2 kg/ha ($5\frac{1}{4}$–$8\frac{1}{4}$ lb/ac), the lower rate if resistant annual weeds only are a problem, can be used as a spot or overall treatment in November or from February to March in occasional years, followed in succeeding years by a cheaper programme.

Creeping buttercup is resistant to all these herbicides except propyzamide at the higher dose. It may be controlled by a directed spray of paraquat applied in March. Bindweeds, creeping thistle and creeping buttercup can be controlled by a directed spray of MCPB at 2·2–3·4 kg/ha (2–3 lb/ac) applied in August or September.

To avoid any adverse effects from residues of herbicides on subsequent crops in the final year before grubbing a plantation do not use chlorthiamid, propyzamide or dichlobenil. Do not apply simazine within seven months of sowing a following crop.

Spray applications

In the past many plantations were carefully sprayed for pest and disease control using high volume rates of over 2200 litre/ha (200 gal/ac) applied through hand lances. This enabled a skilled operator with angled nozzles on the lance to direct the spray up under the bush and through the leaves and shoots, ensuring full coverage by spraying until runoff. Except on a very small scale hand lance spray application is now uneconomic and automatic methods are used. Sprayers can be of two types, air-carrier or hydraulic.

Herbicides are usually applied to the ground using hydraulic sprayers with horizontal booms fitted with fan or cone type nozzles. One feature of all hydraulic pressurised nozzles is the formation of some very fine mist-like droplets at the edges of the nozzle. These droplets are liable to drift on to susceptible crops and possibly cause damage. In general the higher the pressure the more likely will be the formation of these small droplets. For this reason, a nozzle should be chosen which operates efficiently at as low a pressure as possible. A new development which will help overcome the hazard of herbicide drift is the use of the spinning disc applicator. In this technique the herbicide is fed into a spinning disc, and spun off the edge of the disc as droplets of fairly uniform size. The size of the droplets is dependent on two controllable factors, namely the speed of the disc, and viscosity of the herbicide concentrate. In this way it is now possible to apply droplets which are large enough not to drift, and yet small enough to eliminate the need to carry large quantities of water solely in order to apply the herbicide evenly.

Most ground crop sprayers have a fairly small tank, of about 225 litre (50 gal) mounted on the three-point linkage of the tractor. The pump is run off the tractor pto. The capacity of the pump limits the output of these types of sprayers, which are therefore not suitable for applying pesticides and fungicides to established bush fruit.

If a sprayer has to be used both for herbicide and pesticide applications it must be very thoroughly cleaned out after any herbicide application by pumping through clean water plus a suitable detergent and then two or three tankfulls of clean water. There is always a risk of contamination, particularly after the use of MCPA, MCPB, 2,4-D and similar type herbicides.

Susceptibility of some annual weeds to herbicides used in bush fruits

Time of herbicide application	Post-planting, pre-weed emergence						Post-weed emergence		
	Chlorpropham	Diuron/chlorpropham	Chlorthiamid or Dichlobenil	Diuron	Propyzamide	Simazine	Paraquat	Propyzamide 1.7 kg/ha	Propyzamide 3.4 kg/ha
Annual meadow-grass	S	S	S	S	S	S	S	S	S
Black bindweed	S	S	MS	MS	S	MS	S	S/MR	S/MS
Black nightshade	R	R	R	S	S	S	S	S/MR	S/MS
Charlock	MR	MR	MR	MR	R	R	MR	R	MR/R
Cleavers	R	R	S	S	S	R	S	MR	S
Common chickweed	S	S	—	S	S	S	S	S	S
Common fumitory	MS	MS	—	—	MS	MS	S	MR	MS
Common hemp-nettle	S	S	—	MR	R	S	S	R	R
Deadnettle	MR	MR	S	—	R	S	S	R	R
Fat-hen	MS	MR	—	MR	S	MS	S	S/MR	S/MS
Field pansy	R	—	—	—	—	MS	S	—	—
Field penny-cress	S	S	S	—	S	S	S	—	—
Fool's parsley	R	—	—	S	—	S	S	—	—
Groundsel	R	MS	—	MS	R	S	MR	R	R
Knotgrass	S	S	MS	S	S	MR	S	S/MR	S/MS
Maywed	R	MS	S	S	R	S	S	R	R
Parsley piert	R	MS	—	—	—	—	—	—	—
Redshank	MS	MS	MS	S	S	MS	S	S/MR	S/MS
Scarlet pimpernel	R	R	—	S	R	S	S	R	R
Shepherd's-purse	MS	MS	MS	S	S	R	S	MR	MS
Small nettle	S	S	S	MS	S	S	S	S/MR	S/MS
Smooth sowthistle	R	R	MS	MS	R	S	S	R	R
Speedwells	MR	S	MS	R	S	MR	S	S/MR	S/MS
Spurrey	S	S	S	S	—	S	S	—	—
Volunteer cereals	S	S	S	S	S	S	MS	S	S

S = susceptible. MS = moderately susceptible. MR = moderately resistant. R = resistant.

Susceptibility of some perennial weeds to herbicides used in bush fruits

	Aminotriazole	2,4-D or MCPA	Dichlobenil	Diuron	MCPB	Paraquat	Propyzamide
Bindweeds	MR	MS	MS	R	MS	MR	R
Couch/spear grass	S	R	MS	MS	R	MR	S
Creeping thistle	S	MS	S	MR	MS	MR	R
Creeping buttercup	S	S	MR	MS	S	S	S
Dandelion	S	MS	S	R	MS	MR	R
Dock	MS	MR	R	R	MR	MR	MR*
Hogweed	MS	MR	S	R	MR	MR	R
Horsetail	MS	MS	S	R	MS	MR	R
Perennial nettle	MR	MR	S	MS	MR	MR	R

S=susceptible. MS=moderately susceptible. MR=moderately resistant. R=resistant.
*Dock seedlings are controlled at the highest dose.

PEST AND DISEASE CONTROL

There are several serious pests and diseases of black currants that could be present in all commercial plantations and cause serious reduction in growth and hence crop if not controlled. Other diseases and certain pests occur erratically and provided a careful routine check is kept, particularly in the early weeks of the growing period, routine control measures are not always necessary and the specific controls are only applied if necessary.

Thus an efficient spray programme is essential and as no chemical is better than its application it is necessary not only to choose the correct chemicals but also to ensure that they are correctly applied.

The black currant is a difficult spray target. The thick growth and mass of leaves on long slender stalks makes penetration of the spray into the bush and on to both sides of the leaves and flowers difficult, as the pressure of the spray itself tends to blow the leaves so that they mask and protect each other.

HIGH VOLUME HYDRAULIC SPRAYERS

These sprayers rely on the pressure of the hydraulic pump on the spray liquid itself to pump it through suitable nozzles, where it is broken down into droplets which are thus sprayed into the bushes. Hydraulic sprayers for bush fruit will need nozzles that will provide at least 1100–2200 litre/ha (100–200 gal/ac) rate at normal tractor forward speeds.

The tank may be trailed if over 450 litre (100 gal) capacity, or if below this may be mounted on the three-point linkage. Power for the sprayer is provided by the tractor pto, and the sprayer itself may be mounted on the tractor, or on the trailer tank.

There are several models of mounted hydraulic sprayers that have been especially designed for use in bush fruit plantations. The high volume nozzles are positioned on two adjustable booms fixed vertically to the tank, so that the spray is directed outwards on each side as the unit travels down the row. Adjustments of the nozzles, particularly those at the top and bottom, should be carefully made to get the best angles for spraying upwards and downwards into the bushes. There are usually about five to seven nozzles in each boom. When spraying young bushes unwanted top nozzles can be blocked off with blank discs.

If no fixed booms are used for automatic spraying, these hydraulic sprayers have outlets for the attachment of a high pressure hose for hand lance spraying as mentioned above. Hydraulic spraying, whether automatic or with hand lances, using 1100–2200 litre/ha (100–200 gal/ac) requires a large volume of spray liquid, except for small areas, and a lot of time is spent filling and mixing the spray liquid.

These high volume hydraulic sprayers can also be used for the application of herbicides. The booms are remounted in a horizontal position at a suitable height above the ground so that an even spray, with no gaps and no overlaps, is applied between the bush rows. Many herbicides benefit from being applied at 1100 litre/ha (100 gal/ac) rates rather than at the lower rates more usual with ground crop sprayers.

AIR-CARRIER SPRAYERS

These sprayers operate on the principle that an air stream is used to carry the spray liquid to the target. The liquid has to be broken into droplets either by the air flow itself or by hydraulic pressure. The air-carrier sprayers generally have the nozzles and fan mounted on the rear of the tractor worked off the tractor pto and trail a tank for the spray liquid. Some models have the outlets on the trailed tank.

Most large air-stream sprayers are designed to spray top fruit and not all can be positioned and adjusted to give good spraying of bush fruits, even if the output is increased. Because of the closeness of the sprayer to the bushes, the air velocity must be reduced to avoid any phytotoxic effects, particularly on young tender spring leaves near to the nozzles.

MIXING SPRAY CHEMICALS

When hydraulic high volume sprayers are used, or if air-carrier sprayers are used on large areas, considerable volumes of spray liquid will be required. In order to reduce non-spraying time to a minimum a spray tank above tractor height should be centrally positioned within the bush fruit area. This tank should have a 7–10 cm (3–4 in.) outlet so that the machine can be filled rapidly, since filling through a small tap off mains water will take too long.

The spray operative must be given clear instructions, not only on the specific materials to be used per quantity of wash, but also the method of mixing and, if more than one chemical is being applied, the order of mixing. Proper protective clothing (gloves and face shield, or full protective clothing as required by the regulations, see Appendix II) together with adequate clean and correct measuring, weighing and mixing apparatus and containers are needed. All chemicals should be kept locked up when not in use and the Code of Practice followed for the disposal of surplus containers and spray liquid after spraying is finished.

In order to ensure accurate application, not only of pesticides and fungicides but also herbicides, the operator should complete a simple record sheet at the end of each job recording date, amount of chemical(s) used and volume applied. These sheets can be kept in a loose leaf book and thus usefully provide, for each plantation, the annual spray programme.

SPRAY PROGRAMMES

There are several important pests and diseases of black currant and fuller details of their life histories and suitable control measures are given in Leaflets (Appendix I), which are revised more frequently than is possible for this Reference Book.

The most important pest is the black currant gall mite, because of its association with reversion and annual sprays are necessary to ensure continuing freedom from this pest throughout the plantation's life. Aphids and capsids also may occur each year. Care must be taken to avoid applying these insecticidal sprays during the blossoming period or beneficial pollinating insects may be harmed. There are two major diseases of black currants that are found in most plantations and experience has shown routine preventative control methods are generally essential for leaf spot and mildew. Leaf spot is very seasonal in the amount of damage and is encouraged by wet weather; mildew is more rampant under warm dry conditions. In some years grey mould (*Botrytis*) can be troublesome on flowers and also on ripe berries.

When fruit is intended for processing, or is on contract to processors, the kind and amount of spray materials that can be used may be specified by the processors. The fieldsmen should be consulted about permitted programmes. Many pesticides are only approved for use on bush fruits if minimum intervals are observed between the last application and harvesting. These requirements are given in the annually revised booklet *Approved Products for Farmers and Growers* (Appendix I). All chemicals and mixtures of chemicals should only be applied according to the manufacturers' recommendations.

Main pests

Aphids (Leaflet 176)

Several species of aphid occur on currants. Some species, such as the currant-sowthistle aphid (*Hyperomyzus lactucae*) and the red currant blister aphid (*Cryptomyzus ribis*), disperse in the summer to various herbaceous plants, whereas others, such as the permanent currant aphid (*Aphis schneideri*), remain on black currant bushes throughout the year.

The currant-sowthistle aphid, which is the most important species, lays shiny black winter eggs between bud and shoot in the autumn and these hatch at bud burst. Aphid damage, if unchecked, causes distortion and curling of the shoot tips, and puckering and mottling of the foliage.

Control can be achieved by spraying just before first open flower with an organophosphorous insecticide such as chlorpyrifos, demeton-S-methyl, derris, dimethoate, formothion or oxydemeton-methyl. Endosulfan when used against black currant gall mite (see below) will also give some control of aphids.

If thoroughly applied, winter washes effectively control aphid eggs. Use a tar oil spray only when buds are dormant (December to January). Do not spray in windy or frosty weather, or when the bushes are wet.

Black Currant Gall Mite (Leaflet 277)

Mites of this microscopic species (*Cecidophyopsis ribis*) enter buds during the summer where they feed and multiply. Infested buds become swollen and fail to open or develop in the following spring (Plate I). Mites emerge from these buds from the grape stage to June or early July, especially between first open flower and early fruit swell; mites may be carried to fresh sites by insects or air currents. This pest is of main importance as the carrier (vector) of reversion virus disease (page 60) and the economic life of a plantation largely depends on its successful routine control.

Spray fruiting bushes with endosulfan immediately before first open flower and again three weeks later. Thorough spraying and good wetting are essential. Endosulfan will protect bushes from mite attack but will not eradicate existing infestations. Any reverted bushes found should be grubbed promptly and burnt.

Benomyl, although not as effective as endosulfan, will considerably reduce gall mite infestations, and where this material is used as the standard fungicide lightly infested plantations need be sprayed with endosulfan once only, at the end of the blossom period (about four weeks after first open flower).

Non-fruiting bushes and nursery stock should be sprayed with endosulfan in late April, followed by at least two further sprays, with two to three week intervals between sprays.

Black Currant Leaf Midge

The black currant leaf midge (*Dasineura tetensi*) can be of local importance. First generation eggs are laid in early May in the growing tips. The small, white, legless grubs (larvae) feed on the upper surface of young leaves, causing them to become scarred and twisted, and to remain tightly folded with a pronounced upward and inward rolling of the leaf margins. If larvae are still feeding they may be seen by unfurling the distorted leaves. Pupation occurs in the soil. There are three or four generations during the summer but damage by the first is usually most important. Symptoms of reversion disease may be masked in nursery stock infested with leaf midge.

Demeton-S-methyl or dimethoate sprays for aphids will control this pest. Endosulfan, when used against gall mite, will also control leaf midge.

It is particularly important to control this pest during propagation as infested stock cannot be certified and the pest can be dispersed in soil adhering to transplanted bushes.

Black Currant Sawfly (Leaflet 30)

Adults of the black currant sawfly (*Nematus olfaciens*) appear in May and June, laying eggs mainly in the middle of bushes on the underside of leaves. There are two or more overlapping generations in a season. The sawfly caterpillars feed gregariously on the leaves when small, but later spread throughout the bush and can cause considerable defoliation. In severe cases bushes may appear skeletonised, with only stems and main leaf veins remaining intact.

Bushes should be examined during May and June and if sawfly caterpillars are found they should be sprayed with azinphos-methyl plus demeton-S-methyl sulphone or fenitrothion. A second spray about three weeks later may be necessary, since eggs are laid over a period of several weeks. Because of the overlapping generations it is not easy to obtain satisfactory control with one spray. However, control measures should be directed primarily against cater pillars of the first generation. It is important to ensure that the centre of bushes are well sprayed.

Common Green Capsid (Leaflet 154)

The common green capsid (*Lygocoris pabulinus*) overwinters in the egg stage on many shrubs, including currant and gooseberry. Eggs begin to hatch at the early flower stage. Young capsids resemble very active leggy aphids and feed on the shoot tip growth, puncturing the young foliage. Damaged shoots and foliage become distorted and the leaves are marked with ragged brown scars. On currants, this pest is mainly of importance in nursery rows and on young bushes.

For effective control of capsid spray with chlorpyrifos or fenitrothion immediately after flowering. Moderate control of capsid is also given by endosulfan if this is used against gall mite.

Red Spider Mite (Leaflet 226)

The glasshouse red spider mite (*Tetranychus urticae*) is common on many glasshouse and outdoor crops, including currants. Adult females overwinter in the soil or other shelter, and egg laying commences in April. The mites continue to breed and feed, mainly on the underside of the leaves, throughout the summer months. Heavy infestations cause leaf bronzing and premature defoliation, with consequent effect on cropping in the succeeding year. Red spider mite attacks tend to be most serious in hot dry summers.

If infestations warrant control measures spray just after flowering, or just after picking, with chlorpyrifos, demeton-S-methyl, oxydemeton-methyl or tetradifon. The fungicide quinomethionate, if used against powdery mildew, should keep this pest under control.

PESTS OF LESS IMPORTANCE

Currant Clearwing Moth

The currant clearwing moth (*Synanthedon salmachus*) is a local pest of black currant. Red currant and gooseberry may also be attacked. The wasp-like moths fly in sunny weather, laying their eggs in June. Caterpillars feed on the pith of branches from July onwards, becoming full-grown in the following April. Clearwing larvae can kill the branches within which they have been feeding. Where abnormally high levels of this pest occur useful control may be achieved by applying a drenching spray of azinphos-methyl plus demeton-S-methyl sulphone immediately after picking.

Earwigs

Where black currants are harvested mechanically earwigs may occasionally become a problem, since they are caught and deposited in the boxes together with the fruit, but are not blown out by the trash blower. Where experience shows that earwigs are to be expected they may be controlled by spraying bushes with trichlorphon shortly before picking. A minimum interval of two days must be allowed between spraying and harvesting, but processors may require a longer time interval for crops sent for processing. Carbaryl will also control earwigs but must be applied at least six weeks before harvest.

Bud and Leaf Nematode

The bud and leaf nematode (*Aphelenchoides ritzemabosi*) is troublesome on some stocks of black currant. Infestations are worst after wet seasons and show in the spring as brown, slightly swollen buds which fail to grow out, so that lengths of leafless wood appear, often topped by vigorous tip growth. (Bud failure may also be due to other causes and is not in itself evidence of nematode attack.) On a small scale rigorous cutting out of infested shoots is effective in reducing the spread of nematodes, but there is no effective recommendation for chemical control of these pests. Fortunately, stocks of most commercial varieties are not infected with nematode, and the Certification Schemes help to reduce the spread of this pest.

Slugs and Snails (Leaflet 115)

Large populations of snails (*Helix aspersa* and *Cepaea* spp.) may occur in black currant plantations, particularly if weed control is poor. Although snails do not cause direct damage to the crop they roost in the bushes and cause trouble by contaminating consignments of black currants. Snails, like slugs, feed on plant material, mainly at night, and are particularly active in moist conditions. They will seek shelter in fruit trays left standing in plantations prior to picking. Snails hibernate in the soil, frequently in groups, emerging in the spring. Eggs are laid in the soil during the late summer or early autumn. Snails mature in one or two years, depending on the species; adults may live for several seasons.

Control of weeds by the efficient use of herbicides, particularly on headlands, is a sure way of depleting slug and snail populations in black currant plantations. Contamination of fruit containers may be prevented by placing them on a hard standing, or by separating them from direct contact with the soil or weeds with a polythene sheet.

If snails continue to be a problem in weed free plantations growers should consider special measures, such as methiocarb granules used on the headlands or post-picking copper fungicides which may have a long term effect in controlling snail populations; they are certainly worthwhile where infestation levels are high. Headlands, hedges and ditches may also need to be specially treated.

Main diseases

FUNGUS DISEASES

Leaf Spot

The disease, caused by *Pseudopeziza ribis*, often appears in early May, but may not be seen until June or later. The first symptoms can be seen on the upper surface of the leaf as scattered, dark-brown, shiny spots about the size of a pin head; on the lower leaf surface there appear fawn in colour. Individual spots do not increase greatly in size as their extension is limited by the small veins of the leaf. As the disease develops and more infections take place, however, the spots may become so numerous that they coalesce, to cause browning of the whole leaf and, finally, premature leaf fall. Fruit infection may occur, but is not common.

Throughout the season the spots on the leaves produce conidia (or summer spores) which appear as greyish, mucilaginous masses when the leaves are wet. These spores are carried to neighbouring leaves and bushes by splashes of rain and possibly also by crawling insects. At the end of the season when infected leaves have fallen to the ground, the ascospores (or winter spores) are produced. These do not mature until the following spring when they are released from the dead leaves and carried by air currents into the bushes on which, under suitable climatic conditions, they start the next season's infection. In England the relatively mild winters also permit the overwintering of conidia on fallen leaves, so that these spores are also available to initiate infection in the spring.

If leaf spot starts early and progresses rapidly it may cause defoliation so early that the fruit starts to shrivel before it is picked. When, as is more usual, defoliation occurs after picking, but nevertheless prematurely, the effects of the disease may be less obvious. It is after harvest, however, that the fruit buds which will produce the next season's crop are developing and it has been shown that premature defoliation may result in a loss of as much as 60 per cent of the crop in the year after attack.

Varieties vary in susceptibility to the disease, but no variety grown commercially at present shows much resistance. Bushes under two years old are less susceptible to attack than older bushes.

Leaf spot may be controlled by protective and routine spraying of the bushes, commencing at early grape stage and continuing through the growing season. A wide choice of fungicides is available and includes benomyl, carbendazim, chlorothalonil, copper, mancozeb mixtures, thiophanate-methyl and zineb. Some fungicides which are used for the control of American gooseberry mildew (see below) will also give some control of leaf spot, these include drazoxolon and quinomethionate.

There is some evidence that leaf spot tolerance to benzimidazole type fungicides (benomyl, carbendazim and thiophanate-methyl) can occur, which limits the effectiveness of this group of chemicals. Growers should consult with their spray chemical firms' representatives, or ADAS, about spray programmes to avoid the continuous use of these particular chemicals.

Occasionally the fungus *Septoria ribis* also causes spots on black currant leaves. These can be distinguished from those due to *Pseudopeziza* by their fawn centre, larger size and more clearly defined margin. The disease is only rarely severe and in such cases the spray programme outlined above for *Pseudopeziza* should effect a satisfactory control.

American Mildew (Leaflet 273)

In some years, American gooseberry mildew, caused by *Sphaerotheca mors-uvae*, can be widespread and often severe on black currants. The disease attacks the leaves, which become distorted and covered with a white powdery growth, producing conidia which spread the disease during the summer months. As the fungal growth ages, it becomes brown and felted and minute black perithecia can be seen embedded in it. These fruiting bodies carry the fungus over the winter and, in spring, they release air-borne ascospores which, under suitable climatic conditions, start infection on the young leaves. The disease also infects the young shoots and may be responsible for death of buds and consequent yield reduction. On a small scale shoots with dead or heavily infected tips may be pruned out.

Sprays of benomyl, bupirimate, carbendazim, drazoxolon, nitrothal-isopropyl with sulphur, quinomethionate, thiophanate-methyl or sulphur may be used for controlling the disease. Quinomethionate should not be applied to Seabrook's Black or Laxton's Giant as it may cause damage. There is some evidence that mildew tolerance to benzimidazole type fungicides (benomyl, carbendazim and thiophanate-methyl) can occur, which limits the effectiveness of this group of chemicals. Growers should consult with their spray chemical firms' representatives, or ADAS, about spray programmes to avoid the continuous use of these particular chemicals.

The regrowth from stools of bushes cut down for destructive mechanical harvesting, or propagation, is especially susceptible to mildew attack, and fungicidal sprays should be applied to the autumn and spring regrowth at frequent intervals. Shoots from stools used for propagation will need protective spraying from mid-May until the end of August.

Grey Mould

This disease, caused by the fungus *Botrytis cinerea*, can attack the shoots, flowers and fruits. The fungus attacks the shoots causing a die-back of the tips, and frequently attacks the flowers and the fruit. The disease is especially severe after periods of wet weather and frequently attacks the flowers if they have been damaged by frost in the spring and can cause damage to split or over-ripe fruit. The fungus produces its spores in grey-coloured pustules on dead or dying tissues during humid weather and these are disseminated by wind and rain to attack healthy plant parts.

Sprays of benomyl, carbendazim, dichlofluanid or thiophanate-methyl will give control of grey mould and should be applied according to the manufacturers' recommendations. However, strains of *Botrytis cinerea* have been found to be tolerant to benzimidazole fungicides, which include benomyl, carbendazim and thiophanate-methyl; where these strains occur satisfactory control may not be achieved and an alternative fungicide should be used.

Rust

Rust, caused by *Cronartium ribicola*, may be seen on the leaves in some years and may cause early defoliation, but generally it occurs late in the season and no special control measures are warranted. The early stages of attack can be seen as dusty yellow pustules on the undersides of leaves. From these are produced the spores which spread the disease from one bush to another. Later in the season the same pustules produce hair-like curved structures on which are borne a second type of spore; these on germination produce secondary spores which infect Weymouth and other five-needled pines to cause cankers. On these cankers masses of yellow spores, capable of again infecting currants, are produced. Thus the disease is most severe in the neighbourhood of those pines which act as secondary hosts.

Honey Fungus or Armillaria (Leaflet 500)

Black currants may be attacked in the same way as gooseberries by the soil-inhabiting fungus *Armillaria mellea*. Rhizomorphs are produced less frequently than on gooseberry but, underneath the bark of affected roots and in the collar region, the characteristic fan of cream-coloured mycelium can be found on dead and dying bushes. Infected bushes should be dug up and the source of infection, usually an old tree root, should be removed and burnt, leaving in the soil as few roots as possible.

VIRUS DISEASES

Reversion (Leaflet 277)

Reversion, caused by a virus spread by the black currant gall mite (*Cecidophyopsis ribis*), is responsible, more than any other factor, for limiting the economic life of black currant plantations. No variety is immune to this disease. It causes leaf and blossom abnormalities, accompanied by progressive yield reduction. Symptoms do not appear until the year after infection and are at first restricted to a few shoots. Invasion of the bush becomes complete by the third or fourth year, when a virulent strain of the virus will have induced virtual sterility. Once infected there is no recovery.

Blossom symptoms are seen best at late-grape stage, immediately before the first flowers open. Flower buds of healthy bushes are purplish-red, but appear purplish-grey because of a thick down of minute hairs covering their entire surface. By comparison, flower buds of reverted bushes are almost entirely hairless, so that the purplish-red colour is prominent (Plate I). All varieties show this difference clearly, but the normal flowers of some such as Cotswold Cross, Goliath and Baldwin are more hairy and therefore less brightly coloured than in others, for example Wellington XXX, Mendip Cross and

Boskoop Giant. With practice, reverted bushes can be readily picked out, often at an early stage of infection when only single shoots or even individual trusses are affected, but diagnosis should be attempted only when the flowers are dry; the distinction is obscured by wetness immediately after rain or spraying.

With some strains of reversion an early non-recurring symptom is a yellowish line-pattern along the leaf veins, which is inconspicuous and easily overlooked but more likely to develop on basal shoots. There are also characteristic and permanent changes in leaf shape, most readily detected in early summer on fully expanded leaves near the tops of well-grown leader shoots. Reverted leaves have fewer veins, fewer and coarser marginal teeth, a flatter base, a coarser texture and sometimes a darker green colour (Plate VII). The pronounced symptoms are easily recognised but experience is needed to detect incipient reversion from leaf abnormalities.

In a healthy leaf there are five main veins and normally, on either side of the large central lobe, there are between eight and five sub-veins, each arising directly from the midrib and terminating in a marginal tooth; a reverted leaf always has fewer sub-veins and serrations or teeth. The veination and marginal indentation of healthy leaves differ slightly between varieties. In Westwick Choice the teeth on healthy leaves, being coarse and deep, may suggest reversion. On any variety reversion-like symptoms are liable to occur on leaves produced from damaged buds or on forked growth. Such false reversion is usually apparent on leaves at the base of shoots, which in true reversion show little abnormality.

Regular spraying is necessary to control the gall mite (page 55) and hence reversion. Healthy bushes are very resistant to this pest, but their resistance is broken down following infection with the virus, so that reverted bushes are very liable to be attacked and show the characteristic infested big buds. Only healthy, certified stocks should be used for planting new fields, which should be sited as far as possible from old plantations in which there may be sources of both reversion and gall mite. Regular, careful inspections should be made both at blossom time and during June; bushes with any sign of reversion being immediately dug up and burnt to prevent mites spreading infection during spring and early summer when migrating from old galls to buds on new growth.

MINOR VIRUS DISEASES

These diseases would be much more serious if more common. Every care must be taken to avoid spreading them through propagation from bushes showing any of the symptoms described below.

The aphid-transmitted gooseberry vein-banding virus occasionally infects black currants, causing slight stunting but little crop reduction. Sensitive varieties develop broad yellow vein-banding of the first leaves to expand in spring, followed by similar, though usually less obvious, symptoms. Entire leaves of Mendip Cross may develop a vein-net pattern but in other varieties, such as Amos Black, vein-banding is normally restricted to individual lobes of occasional leaves that become slightly distorted and asymmetrical; Baldwin and Wellington XXX seldom show any symptoms. Virus-induced vein-banding is more clear-cut than the more diffuse mottling caused by aphid feeding and is present before aphid infestations develop.

Green mottle is caused by the widespread cucumber mosaic virus, which is transmitted by aphids to many crop plants and common weeds. The virus does not appear to spread readily into black currants; infected bushes, though stunted and yielding little crop, have been found only occasionally in commercial plantations. Foliage symptoms are variable and best seen as the leaves become fully expanded, tending to be inconspicuous in immature or old leaves. Large sectors of some leaves become pale green, or discolouration is restricted to broad bands along certain main veins, sometimes giving a 'watermark' effect best seen by viewing the leaf against the light.

Yellow mottle is caused by arabis mosaic virus, which is transmitted by the soil-living nematode *Xiphinema diversicaudatum* and infects various crops and weeds. Infected bushes give little fruit but infection in black currants is rare. The first leaves to expand in spring show a conspicuous yellow mottling, irregularly distributed or forming spots and rings. Symptoms are less conspicuous on leaves of extension growth and are barely detectable by mid-summer. Stock plants should not be raised on land where the virus and its nematode vector (carrier) occur.

Black currant yellows is very rare, though the symptoms may have been confused with those of nitrogen or other nutrient deficiencies. Infected plants lack vigour and give a reduced crop. Indistinct yellowish flecks appear on the foliage in spring, followed by a more distinct olive-green discolouration affecting large sectors of the leaves, and symptoms are more conspicuous after warm sunny spells. The disease spreads slowly in the plantation but by what means is not known.

Infectious variegation (gold dusting) occurs throughout the stocks of a few little-grown varieties, such as Daniel's September, which develop a bright chrome or pale-yellow mosaic pattern on the early leaves, followed in summer by a broad yellow banding of the main veins. Graft-inoculated Baldwin and Wellington XXX proved to be tolerant, developing only very slight symptoms in occasional seasons. It is not known whether the disease can spread naturally.

Harvesting

Black currants have traditionally been picked by hand, payment being by piecework. Unlike most other fruit crops the size of the individual fruits is generally not important for processing but the fresh market requires good bold consignments. Provided all the currants are firm and none is still green, the considerations for efficient marketing are to get the fruit picked as quickly and cheaply as possible without crushing the berries, free of debris and with the minimum of dropped and wasted berries.

The ability to obtain gangs of casual pickers has often limited the growing of black currants in areas otherwise suitable and inadequate labour has frequently resulted in good crops being badly harvested and not meeting the processors' quality standards. The increased efficiency of weed control with herbicides means that hand harvesting has now become the biggest single cost in the production of black currants, often as much as two-thirds of the total annual costs.

Because of the difficulties in harvesting this crop considerable research and development work was done to produce satisfactory mechanical picking machines and several types are now available. A destructive harvesting system was used in the early 1960s and harvesting from the standing bush since the early 1970s. The area being mechanically harvested has increased each year and now most of the crop is harvested in this way, only very small areas being still picked by hand.

HAND PICKING

To ensure an adequate number of pickers growers may have to advertise in urban areas and hire transport to bring the pickers to the fields. Care must be taken that adequate insurance has been obtained to cover all eventualities. These overhead costs often amount to an additional 10–20 per cent of the actual piecework rate.

It is considered that there should be at least one supervisor to each 60 pickers in addition to the staff check-weighing the fruit and filling and stacking trays. On most holdings this means that the regular workers have to undertake supervisory duties during the harvesting season and they would benefit from some help and instruction beforehand on how best to handle large numbers of casual workers. The Agricultural Training Board can help in arranging suitable courses of instruction for supervisors in local areas on request.

Depending on the crop yield considerable gangs of workers, usually women and children, are required per area of bushes. It is generally considered that a really experienced hand picker working on well grown bushes carrying a good crop may be able to harvest about 50 kg (110 lb) per day. But the picking rate varies widely and many casual workers, particularly those accompanied by small children, will often only work a very short day, and not achieve half this output.

Fruit that is firm and dry is much easier to pick than that becoming over-ripe. Also there are often spells of wet weather during the black currant harvesting period of late July to early August and this will hold up hand picking, so that for any one variety it may well prove there are barely 10 days in which to pick the crop. Thus about two skilled, or at least four unskilled pickers working for 10 full days are needed to pick a tonne (ton) of fruit.

If the fruit is for processing a large gang of pickers is required, so that adequate lorry loads can be collected each day, or at the latest every other day. The fruit, once picked, needs to be removed from the field and dealt with as rapidly as possible, to avoid overheating, spread of rots and deterioration. Thus the picking gang will need to be big enough to harvest at least 3 tonnes (3 ton) or more to ensure a lorry load can be collected each night, and this requires at least 60 experienced pickers, or two or three times this number if they are inexperienced.

It is essential that all the currants for processing are coloured and none still green. Picking has to be delayed until this stage is reached, since the crop is cleared at one pick, and although a few green, unripe strigs can be left in the centre of the bush if necessary, the chief problem is to delay harvest until all the fruit is adequately ripe, without the risk of inadequate time and labour to clear the crop before it becomes over-ripe and drops.

ORGANISATION OF PICKING

Careful planning beforehand will ensure that pickers are clearly shown their positions in the plantations and start work without delay. Adequate supervision of pickers is essential to ensure:

1. Fruit is not crushed.
2. Bushes are properly cleared.
3. Unreasonable amounts of berries are not dropped on the ground.
4. Leaves and other trash are kept out of the containers.

The fruit is picked into plastic buckets holding 5–6 kg (11–13 lb) fruit. These buckets should have a hole drilled in the base so there is less temptation of theft. Two buckets are provided to each picker.

Inexperienced pickers must be shown how to position the bucket below the outmost branches and how to nip off each strig to avoid crushing the basal berry. Both hands should be used and if necessary the end of a branch can be tucked and held under the armpit once the fruit is cleared from the tip, thus aiding fast picking. As the outer branches are cleared the bucket is moved nearer the centre of the bush, the branches parted and the inner fruit picked.

Where rows are continuous pickers should not be allowed to force their way through the rows but instructed to clear their side of the row. Ideally they should work in pairs opposite each other and clear each bush, or section, before moving on. This is seldom practicable and adequate supervision is needed to ensure inner branches of fruit are not left unpicked. Naturally workers on piecework tend to pick the easy fruit and leave that requiring more effort. Before starting the pickers can be shown two buckets with the required weight of fruit in each; this is helpful, as they will then not overfill the buckets before going to the check point and checker.

CHECK POINT

The check point needs scales, the counterweight type are best with a bucket and the required weight on the one side. The incoming buckets are check weighed, excess fruit gently scooped off into a spare bucket which the picker then takes away with another empty bucket. On some holdings pickers work with four buckets but spillage can result.

The checker may pay cash for each bucket or tray filled, or give tokens which can be changed for cash at the end of the day, or the fruit may be booked against each picker's name. Care must be taken to ensure adequate reliable staff at the checkpoint to avoid queuing. The checkpoints should not be too far away from the area being picked, 70–90 m (75–100 yd) is a maximum, or time will be wasted walking to and from the point.

PROCESSING TRAYS

Currants for processing are poured into wooden trays holding 10 kg (22 lb). The tray has to be lined with a sheet of polythene film to enable the frozen block of fruit to be turned out after freezing. This liner also keeps the fruit clean and holds any juice. The trays should be stacked carefully on delivery either on hard standing or on polythene sheets to prevent soil adhering and avoid slugs and snails. Muddy trays will drop mud on the full tray below when stacked. Empty trays and the liners are taken out into the field on the tractor trailer and stacked as filled. Where possible the trays should be stacked on to pallets so that a tractor with rear loader, or if conditions are suitable, a fork lift truck can be used to load the lorry each evening. If possible the full trays should be moved to a shady position on a hard standing area for grouping to make up the lorry load.

MARKET CONTAINERS

To avoid rehandling, fruit to be sent to the wholesale market is preferably picked direct into market containers. Very good quality or early fruit may be marketed in 0·5 kg (1 lb) punnets within outer containers.* Alternatively non-returnable trays (tomato type) or fibreboard baskets may be used. Care must be taken to keep containers clean when taken into the field.

When picking for market the pickers must be extra careful not to crush or 'milk' the berries and the crop must be very firm. Fruit that is over-ripe will be crushed in transit. Very early varieties such as Tor Cross, Mendip Cross or the earliest ripe fruit on French Black or Wellington XXX may repay picking over at time rates with a bonus for fast work so as to ensure a quality sample in punnets and a commensurate return from the market.

MECHANICAL HARVESTING

Destructive harvester†

In this system the bushes are partly cut down, the fruiting branches being fed through a stripper mechanism to remove the berries. The system most commonly adopted uses a static stripping machine on the headland. The fruiting branches are cut by hand and transported to this machine. To reduce the labour required in transporting and feeding the branches into the stripper the 'carpet' haulage system was developed. In this the branches, which are cut by hand loppers or pneumatic pruners, are placed on strong woven nylon sheets which are laid at appropriate intervals between the rows of bushes. When a sheet is fully loaded it is winched on to a special transporter incorporating a conveyer floor. The transporter is tractor hauled to the stripping machine and with the aid of the 'carpet' the branches are mechanically fed into the machine with assistance from workers. To obtain maximum efficiency the positioning and frequency of moving the stripping machine have to be carefully considered. The 'carpet' haulage system requires a team of about 10 workers who should be able to harvest about 7 to 10 ha (18–25 ac) per season. After drying the branches can be burnt on the headland and this avoids the need for removal of winter prunings from the row.

In the past efforts were made to cut off the whole bush and feed this to the machine, the aim being that the plantation would be cropped every other year

---

*There is a British Standards Institution standard for wooden trays (BS 2892: 1974) and for non-returnable fibreboard trays (BS 3789: 1975) obtainable from the BSI, 101 Pentonville Road, London N1 9ND.
†This machine is no longer in production but will be made to order.

only on the two-year old shoots. These methods were unsuccessful as the very vigorous growth from the cut down bushes led to many problems.

The present selective pruning method eliminates winter pruning, as the fruiting branches are carefully cut out. However, it is usual for growers still using this static headland harvester to prune out only one third of each bush for mechanical stripping and to hand pick the remainder. The system works best with young bushes with heavy crops so that the minimum of branches have to be handled.

This headland harvester is no longer in production but machines are still being used on those farms where a successful method has been evolved. The amount of labour required, though less than for hand picking, is much higher than with mobile picking machines. Compared with mobile harvesting from standing bushes, this system permits better inspection of the fruit as the trays are filled and the rather heavy pruning required ensures that the bushes are kept vigorous and well supplied with new shoot growth.

Harvesting from standing bushes

Besides saving labour this method has the advantage that the bushes remain intact to crop again in succeeding years. Mechanical harvesting can be divided into three main processes, removal of the currants from the branches, separation of the fruit from the trash—mainly leaves—and conveying of the fruit to the trays. The forces required to remove currants from the strigs, or strigs from the branches, are similar to those which cause loss of leaf, bud and shoot damage. Loss of leaf during the harvesting period must be kept to a minimum, since it has been shown that excessive leaf loss at harvest time results in a reduction in the next crop.

Two mechanisms, plucker belts and vibrating fingers, are used to remove currants from the bushes. To minimise the loss of fruit from branches on the outsides of the rows projecting parts of machines, and especially tractors, should be well guarded. As more experience of mechanical harvesting is obtained it is probable that the present machines will continue to be modified and improved.

*Plucker belt machine**

Initially this machine was developed to harvest fruit on the younger branches left in the bushes after selective removal of fruiting wood with the 'carpet' haulage system and static headland machine.

Subsequently this machine was developed into a complete harvester. The unit is designed for mounting on to the 3-point linkage of medium powered tractors and is chain driven via the pto or hydraulic pto pump. The unit is arranged to give the driver a good view of the branches entering the machine and this facilitates accurate steering along the rows. The tractor is driven so that all the branches of the bushes in one row are crowded by guides into a narrow throat which is inclined away from the tractor. On either side of the throat there is a plucker belt, incorporating projecting wire loops, which moving in an upwards direction strip the fruit from the branches. The fruit and the trash from both

---

*No longer in production, being superseded by straddle harvesters.

plucker belts are passed through a high-speed air stream. The leaves are blown away behind the machine while the fruit falls into trays carried at the rear of the tractor. A skid is towed to take a small quantity of full trays.

This type of machine is less costly than those incorporating the vibrating-finger mechanism, but difficulties may occur in feeding large dense bushes into the throat of this machine. Also unless the machine is carefully set the action of the plucker mechanism can be severe. Care is needed during operation to avoid excessive leaf removal and/or berry damage.

Vibrating-finger machines

At present this mechanism of fruit removal is used in a tractor semi-mounted machine and in two self-propelled harvesters.

Semi-mounted harvester. This machine was originally developed by the National Institute of Agricultural Engineering and harvests fruit from half of a row at a time. A crop divider splits the bush and guides the branches nearest to the machine into the shaker mechanism. This consists of two freely rotating cylinders about 150 mm (6 in.) in diameter mounted one behind the other in the direction of machine travel and with their axes at right angles to the row and 45° to the horizontal. Six or seven rings of tines are fitted along each cylinder. The cylinders and tines are caused to vibrate by out-of-balance rotating weights (inertia shakers) positioned at the top of each cylinder. As branches are fed into the shakers they are caught between the tines and vibrated. Normally the shakers are set to vibrate at just over 1000 cycles per minute but this can be adjusted for each cylinder independently over a fairly wide range. The berries and trash shaken from the bushes falls on to inclined cross conveyors which feed a longitudinal belt conveyor. An air blast is used to separate the berries from trash as the fruit is transferred to the longitudinal conveyor from which it is discharged into trays at the rear of the machine. Empty and full trays are carried on a sledge which is towed behind the machine.

The main components of the machine are driven by individual hydraulic motors, which are supplied with oil from a pto driven pump. This system gives the operator fine control of the speed of the shaker units, the conveyors and the fan, independently of each other. Thus the machine can be set for different conditions so as to harvest the maximum amount of crop with the minimum amount of damage.

In addition to the tractor driver this harvester requires a skilled operative who rides on the rear of the machine to continually check the machine adjustments, forward speed of working, fruit removal and bush damage, if any. He is assisted by a team which will vary according to the weight of crop and the handling system adopted. Usually two other workers are required to handle the full and empty trays on the skid behind the machine. Further workers with tractor and trailer, or rear loader, are needed on the headlands to off-load, weigh and stack the full trays and position the spare skids with empty trays. More workers may be required to feed branches into the machine, or if the skids have to be manually unloaded within the row.

Self-propelled harvesters. Both types of self-propelled harvesting machines straddle a row of bushes and harvest most of the fruit from the row at a single pass. As the machines move forward the bushes are split with a divider and the branches inclined outwards. Branches from each half of the bush are fed into one of two shaker mechanisms arranged, when viewed from the front, in V formation.

The shaking units are basically similar to those described in the previous section. Berries and trash shaken from the bushes are elevated to the rear of the machine where the trash is removed by two fans and the fruit discharged into trays. Empty trays are carried on the machine and full ones loaded into two trailers, which are pulled behind the machine with one trailer on each side of the row being harvested. Each trailer will carry about 0·5 tonne ($\frac{1}{2}$ ton) of fruit with some extra capacity provided on the machine itself. When the machines reach the end of the row the loaded trailers are unhitched, the machine turned into the next row and two empty trailers hitched up. A larger, single trailer that is towed over the row of picked bushes is also available.

Self-propelled harvesters are powered by a diesel engine or a built-in tractor unit. These units drive hydraulic pumps which in turn serve hydraulic motors on the main machine components. As previously described this arrangement gives the operator infinitely variable and independent speed adjustment of each component so as to obtain the best results under different conditions.

The driver, who sits above the row being picked, is the key operative and controls not only the forward speed but the rate of shaking and all the mechanisms. He is assisted by workers who ride at the rear of the machine on each side of the row being harvested. These workers take the full trays off the machine as they are filled and stack them on the two trailers being towed behind and place the empty trays with the polythene liner in place to catch the fruit. The number of workers to keep pace with the flow of currants will depend mostly on the yield, since in heavy crops it is the speed that the workers can handle these trays that controls the output. With crops up to about 5 tonne/ha (2 ton/ac) one man on each trailer can cope, but with increasing yield the team has to be increased, a total of four workers being required when conditions and yield are good. Additional staff are required on the headlands to unload the trailers, weigh off and stack the trays and reload the empty boxes on the harvester as it turns.

Machine operation

The amount of fruit left unharvested on the bushes, or waste berries dropped on the ground, is generally considered to be no more than that experienced with hand picking.

Under normal conditions it is reasonable to expect that the available machines will be able to harvest 80 per cent or more of the crop, with maximum efficiency in plantations especially planted for machine harvesting (page 29). With existing plantations which were planted for hand picking, it is possible to improve the proportion of crop picked by:

(a) guiding branches by hand into the machine so as to prevent bunching of the branches;

(b) going over the row again, preferably in the opposite direction, either immediately after the first pass, or after a few days;

(c) the use of chemicals which induce abscission. In many cases there are difficulties in getting spraying machines to apply these chemicals through the plantation just before harvest.

Ethrel is one such chemical which by the liberation of ethylene hastens ripening and induces fruit abscission. However this chemical is expensive and the chief uses may be to hasten the ripening of part of a large area of Baldwin black currants so as to permit an earlier start to be made than if all the fruit ripened naturally, or to allow calibration and adjustment of the harvester for a day or so before the main season starts. After using a machine on ethrel sprayed bushes it is important to readjust before the machine is used on those not sprayed.

None of the mobile machines will harvest fruit situated near the ground, generally only berries 250 mm (10 in.) or more from the ground can be caught on the belts. Thus the machines are not worth using over bushes in their first cropping year and, if growth is rather compact, even two-year-old bushes may not be suitable for machine harvesting.

Machines catch and retain trash similar to fruit, also earwigs, snails, small knobs of wood and dead shoot pieces, soil and stones. The rate at which fruit is delivered into the tray prevents all but a cursory examination of the fruit. Plantations where slugs, snails or earwigs are troublesome will need spray and other treatments to control these pests. Hand pickers tend to dislodge pests and most fall onto the ground rather than in the bucket, but with mechanical harvesting a bigger proportion are caught on the belts. When bushes are first shaken the accumulation of loose shoots and trash is dislodged and generally there will be less trouble from this cause in subsequent years. If unreasonable quantities of soil and stones are being collected the machine may need adjustment. An even soil surface beneath the bushes aids mechanical harvesting and, particularly with straddle machines, straw mulching should not be used as it tends to ball up on the catching belts.

It is essential to clean harvesters at least once per day, washing off all the crushed berries, trash and juice from the conveyor belts and shaking or plucking mechanisms. As soon as the harvest is finished the machine should be cleaned thoroughly, serviced according to manufacturers' recommendations and any worn or suspect parts replaced or the parts ordered.

Machine choice

Both the straddle and tractor shaker harvester are expensive machines so that for minimum costs per tonne harvested they should be used to maximum effect each season. Fruit can be harvested at any time provided the berries are reasonably dry. Shift working, with a bonus per tray harvested shared between each team, has been found by some growers to help in the operation. Although most of the work only requires strong unskilled labour it is essential that an experienced man, capable of adjusting the machine and interested in its proper utilisation, should always be in charge.

The area which can be harvested in a season depends upon the planned working time (hours/day × number of working days) and the rate of harvesting. Due allowance must be made in the planned working time for unavoidable stoppages, for example due to unsatisfactory conditions or failure of part of the system.

The rate of work depends upon many factors including yield, type of bushes, layout of the plantation and number and quality of workers employed. Thus it is not possible to give more than a general indication of the overall rate of work which can be expected from different types of harvester. Until more precise information is available it is considered reasonable to plan on the basis that a self-propelled straddle machine will harvest at the rate of about one tonne/hr (one ton/hr) while semi-mounted single sided machines will work at about half this rate. Before applying these figures to individual cases it may be necessary to modify them considerably to take account of different situations.

Where the area and yield of currants on an individual farm does not justify the purchase of a harvester the formation of a syndicate should be considered. It is sometimes possible to arrange for the contract harvesting of black currants; generally the contractor provides the machine and skilled operator, the other labour being provided by the grower.

Economics

The assessment of the potential profitability of a new black currant enterprise can be a difficult problem. This is because, in addition to the normal risks of fruit growing, such as irregular cropping and a variable market price, the black currant is a perennial crop that takes some years to come into production but is then expected to remain viable for 10 years or more. The cash flow is therefore irregular and the time when income is generated becomes all important since the initial expenditure on labour, materials and specialised equipment will not give any real return for two or three years. Apart from cash flows other factors must be considered. Will the existing labour force be large enough to cope with the new enterprise and are the necessary skills available? Will sufficient casual labour be available to cope with harvesting, or how much extra capital must be borrowed or set aside to buy a mechanical harvester? Will other specialised equipment, for example a sprayer and fertiliser spreader, have to be purchased?

A thorough investigation of the sales outlets is essential. Markets for fresh fruit are limited and where the crop is to be grown for processing the negotiation of a long-term contract has obvious advantages. It must be recognised that black currant growing has never been a stable industry and prices may fluctuate widely from year to year.

THE PHYSICAL DATA

The first step is to assemble all the available information. Tables 1 and 2 give examples of possible yield patterns and the price structure that has prevailed over the past decade. Tables 3 and 4 set out the materials that will be required to establish an area of black currants and Table 5 lists the approximate times the main cultural operations could take and provides information on the labour requirements in a typically fully cropping year. Naturally in practice there will be wide variations from these suggested figures and each case must be assessed on its merits. Columns have been left blank so that current costs can be included for any particular year. This physical information can then be used to help draw up a budget for the new enterprise.

Table 1 Approximate yields

Site/soil	Yield tonne/ha			Yield ton/ac		
	Low	Medium	High	Low	Medium	High
Good	5·0	7·5	13·8	2·0	3·0	5·5
Average	3·1	6·3	8·8	1·25	2·5	3·5
Poor	1·3	4·4	5·6	0·5	1·75	2·25

Yield after planting

Year	1	2	3	4
Average t/ha	0	2·5	5·0	6·3
Average ton/ac	0	1·0	2·0	2·5

Table 2 Average Processing price/tonne (£)

1971	1972	1973	1974	1975	1976	1977	1978	1979
170	150	220	250	187	295	876*	520	410

*Frost year.

Table 3 Materials to establish an area
Planting 0·6 × 2·7 m (2 × 9 ft) or cuttings 0·3 × 2·7 m (1 × 9 ft)

	Per hectare	Cost £	Per acre	Cost £
Bushes	6180		2420	
Cuttings	12 350		4840	
Farmyard manure	50 tonne		20 ton	
Fertilisers				
Compound 15: 15: 20	400 kg		3 cwt	
Nitrogen (34·8%)	250 kg		2 cwt	
Herbicide				
Simazine cp	5 kg		4 lb	
Sprays				
Insecticide (×1)				
Fungicide (×4)				

Table 4 Approximate labour and machinery time to establish an area

Tractor and operator	Hours/ha	Hours/ac	Cost £
Ploughing and cultivation	10	4	
Fertiliser pre-planting	2·5	1	
Bulky manures	10	4	
Mark out	2·5	1	
Plough-out, to plant and back	15	6	
Cultivation to level	5	2	
One spring fertiliser	2·5	1	
One spring herbicide	2·5	1	
Five sprays for pests and diseases	12	5	
Labour			
Planting and cutting down	64	26	
Spot treating weeds	5	2	

Table 5 Typical annual labour requirements for an established plantation (per hectare and per acre)

	Hours/ha	Hours/ac	Cost £
Fertiliser			
+application (tractor)	2·5	1	
Herbicide(s)			
+application (tractor) × 1–2	2·5–5	1–2	
+spot application (hand)	5–12	2–5	
Insecticides × 2–3			
+application (applied with fungicides)			
Fungicides × 4–8			
+applications (tractor and sprayer)	10–20	4–8	
Pruning, labour only (casual)	62	25	
Pruning, clearing or pulverising (tractor)	10	4	
Harvesting (see page 70)			

GROSS MARGIN ANALYSIS

One budgeting approach is to use the Gross Margin technique to provide a simple first appraisal of the proposed new enterprise. Using the Gross Margin method the costs of production are considered separately as Variable or Fixed Costs. Variable Costs are those which arise directly from the new enterprise. Thus the main Variable Costs incurred in establishing black currants will be the bushes, herbicides, sprays and other materials and additional casual labour employed.

Fixed Costs are defined as those which are common to all the crops grown on the holding. Thus regular labour, farm machinery, rent, rates and other overheads are generally treated as Fixed Costs.

The allocation of Fixed and Variable Costs will vary from farm to farm, but provided this fact is recognised and it is fully appreciated that the Gross Margin is indeed only a margin and by no means a measure of net profit, the system offers a good method of assessing the viability of black currant production.

Because of the variable cash flow pattern it is convenient to study the project in several phases:

(a) The initial planting and establishment period when there will be a heavy commitment to purchase bushes and materials, pay casual labour and bring the bushes into bearing.

(b) A period when the bushes start cropping but are still young and only a small income can be expected. Expenditure on specialist equipment may be necessary at this stage.

(c) A longer period at full production.

(d) The final expense of grubbing the plantation.

The establishment of black currant plantation will require the injection of considerable capital resources which must be written off during the life of the crop. The income from the fully cropping enterprise should be sufficient therefore to cover this item of capital repayment as well as interest on the capital invested, before it really begins to contribute to the general farm income. All costs incurred before the black currants are established may be considered as a mortgage or loan and the effect of writing off this sum studied at an appropriate rate of interest over the expected life of the plantation.

By consulting Amortisation Tables of the type used by building societies, which calculate the sum required each year to pay interest on a loan and to repay the principal involved, the annual capital charge for the project can be calculated. Another approach for a long-term crop such as black currants would be to use the principle of Discounted Cash Flows, which gives due weight to the time factor when assessing the value of future income. A full explanation of Discounted Cash Flow budgeting is beyond the scope of this Reference Book.

Problems of capital will vary greatly from farm to farm. For example the financial implications for anyone embarking on black currant growing for the first time may differ markedly from someone who is merely adding to an existing enterprise. Because problems of capital are so variable they have been omitted from the budgeting data discussed here. They are, however, a vital aspect of budgeting and must always be taken into account.

The following calculations are given as an example of using the Gross Margin technique. Since every new black currant enterprise will present a separate and distinct problem, *the figures should be interpreted with the utmost caution.* In every case a prospective grower must substitute his own figures if he wishes to obtain a realistic picture of his own unique position.

Gross Margin budgeting (Example only—1980 prices)

Variety:	Baldwin, planted as two year bushes at $2 \cdot 7 \times 0 \cdot 6$ m (9×2 ft). Weed control with herbicides.
Life of plantation:	12 years for processing.
Estimated total crop:	20 tonne over 12 years. Annual crop range 0–10 tonne/ha (0–4 ton/ac).

1. *The Variable Costs of establishment.* First Year. £/hectare (1980 prices)

Bushes 6180	1265
Farmyard manure	150
Fertilisers	67
Pest and disease sprays	123
Herbicides	76
Extra casual labour for planting and pruning: woman 60 hours at £1·95/hr	117
Total Variable Costs, first year	£1798

During the second year further Variable Costs will be incurred and little if any income produced. In the example these extra charges might bring the total Variable Costs of establishment up to £2000/hectare.

2. *The Variable Costs of production when fully bearing.* £/hectare (1980 prices)

Fertilisers	53
Pest and disease sprays	227
Herbicides	82
Casual labour 86 hours at £1·95/hr	168
Total Variable Costs of production	£530

Using these data a gross margin can be calculated for each cropping year; an example is given here for the first five years.

Variable Costs and Gross Margins year 1–5 (per hectare)
Based on 10 ha (24 ac) unit for processing fruit picked by straddle type harvester (1980 prices)

Year	1	2	3	4	5
Yield (tonne)/ha	0	1·0	5·0	5·0	6·3
£/tonne	410	410	410	410	410
Crop output £/ha	0	410	2050	2050	2583
Variable Costs					
Bushes	1265	0	0	0	0
Materials	416	362	362	362	362
Casual labour—growing	117	168	168	168	168
Casual labour—harvesting	0	165*	123	123	123
Total Variable Costs	1798	695	653	653	653
Gross Margin	−1798	−285	+1397	+1397	+1930

*Hand picking costs.

The limitations of a simple Gross Margin analysis cannot be overemphasised. A much fuller picture of the financial implications may be obtained if a complete cash flow is calculated to cover the expected life of the plantation. This can take account of the individual grower's precise financial situation and must assume some knowledge of forward prices. In such an analysis an appropriate allocation of the Fixed Costs may be included, particularly the purchase of any specialised machinery, including for example the straddle harvester implied in the Gross Margin calculation shown above. Farm Management advisers are available to help in the use of such techniques.

HARVESTING

Harvesting is the major annual cost when the plantation is in production. If harvesting by hand the total cost is then directly related to the yield. Ten per cent on the total piecework costs should be allowed to cover lining trays, stacking, carting from the field and loading onto the processors' lorries. An additional 10 per cent can be allowed to cover supervision of pickers in the field, insurance, pickers' facilities, picking buckets and other incidentals, including some advertising.

If black currants are being hand picked to be sold on wholesale markets in non-returnable wooden trays or fibreboard containers, the cost of the containers must be allowed for as well as extra labour costs for weighing and checking and costs of transport to the market and salesmen's commission. It may also be necessary to pay a bonus over normal piecework rates to ensure more carefully picked fruit.

MECHANICAL HARVESTING

Nearly all large areas of black currants for processing are now picked by machine. The merits of mechanical harvesting and the choice of the best machine are problems that must be separately assessed for each particular situation.

For straddle machines costing about £30,000 the annual costs over five years, including depreciation, capital repayment and interest, repairs and maintenance, are about £8 520. Seven casual workers are required per machine. The work rate of the machine is about 7·5 hours per hectare (3 h/ac).

Calculated cost 1980 (£) of straddle machine harvesting at different yields from various areas

Cost (£/tonne)					Cost (£/ton)				
yield/hectare (tonnes)	2·5	5·0	7·5	10·0	yield/acre (tons)	1	2	3	4
area harvested (hectares)					area harvested (acres)				
40	336	67	45	34	100	134	67	45	34
30	407	81	54	41	75	163	82	54	41
20	549	110	73	55	50	219	110	73	55
10	975	195	130	98	25	390	195	130	98

The table clearly shows that to get low mechanical harvesting costs the maximum area of heavy crops should be picked. The calculation also indicates that a machine must handle over 100 tonne (ton), preferably more, per season to be economic, compared to existing total hand picking costs of some £185/tonne. The table is only a guide; similar tables can be readily calculated for different machine and labour costs and work rates.

Gooseberries

The gooseberry is not a native of this country; it was probably introduced from central Europe during the sixteenth century. Until comparatively recently it was an important soft fruit and in 1929 no less than 7690 ha (19 000 ac) were devoted to its cultivation. Since then its popularity has steadily declined and in 1978 only 945 ha (2335 ac) were grown.

Areas, estimated yields and total output of gooseberries
England and Wales

	Area (hectares)	Yield (tonnes/hectare)	Output (tonnes)
1950–1959 (average)	2488	5·57	13 800
1960–1969 (average)	2018	6·12	12 300
1970–1974 (average)	1407	6·93	9700
1975	1087	5·93	6400
1976	1015	6·31	6300
1977	990	4·34	4300
1978	945	5·71	5400

In the past, in addition to its value as a soft fruit, it was used as a major source of pectin by the confectionery and jam industries. This outlet is now closed since better sources of pectin are available, and the crop is now sold either for canning or freezing, or as fresh fruit for culinary purposes. The jam industry takes a little fruit and a small area is still devoted to the production of special dessert berries.

About half the production area is around Wisbech, comprising west Norfolk, the Isle of Ely and Holland (Lincs). The remaining areas are in Kent, Worcestershire, Staffordshire and Gloucestershire. A small area around Chailey and Newick in east Sussex still specialises in growing the variety Leveller as a high quality dessert fruit for the London market.

The gooseberry is the first hardy fruit to yield in the year and in very favoured localities it is possible to pick young green berries by early May. Until late July fruit is sent steadily to the fresh fruit market and the season ends with the marketing of small quantities of ripe dessert berries. The bulk of the crop, particularly from the Wisbech and Midlands areas, is grown for processing, the fruits being harvested at one picking while the berries are still hard and green.

The life of a gooseberry plantation is usually considerably longer than that of black currant plantations and, whilst occasional bushes may be killed by mechanical damage or *Botrytis*, replanting within the plantation is often done with success. There are many old plantations still fruiting well after 20–30 years. The bushes yield a little fruit within a year or two of planting but maximum yields are not usually obtained until about the fifth or sixth year, or later if growth is slow.

Yields vary considerably with growth and season, variety, and stage of maturity at picking time. Under good conditions an established plantation of Careless can yield regularly up to 12–15 t/ha (5–6 ton/ac), picked green for processing. Much lighter crops are obtained from Leveller grown for dessert fruit.

Gooseberries may be grown alone or as an intercrop under fruit trees, although they are undoubtedly easier to manage when grown by themselves. Intercropping is still quite common, particularly in the Wisbech area, where their use as an undercrop in plum or culinary apple orchards is normal practice. They are often grown under plums in the Midlands. The advantages of such a system lie in the income given by the gooseberries whilst the top fruit is coming into bearing and the additional shelter provided by the trees which assists cropping. The disadvantages from intercropping are mainly problems with spraying and picking. As the trees mature the yield from the gooseberries is naturally curtailed, but the bushes will continue to yield even when in the shade of established plums or culinary apples, provided the soil is good and freedom from weeds assured.

One useful point in the management of gooseberries is their flexibility in picking and marketing, since the fruit ripens very slowly. Thus strawberry growers often find it useful to have an area of gooseberries to pick, since it enables them to organise picking gangs early in the year. Also during the strawberry season it is often convenient to switch from strawberry to gooseberry picking and *vice versa* as market and weather conditions change. Gooseberries can be picked under damp or wet conditions when strawberry picking would be unwise.

Birds can be a severe hazard to gooseberry production. The worst culprit is the bullfinch which eats the buds in the winter, often starting to attack the bushes immediately after leaf fall. This not only destroys the potential crop, but it leaves bare branches which take several years to refurbish with fruiting shoots. After a severe attack the bushes may be so damaged that they are hardly worth retaining unless they are cut back to develop new fruiting branches. Sparrows can also cause damage by attacking the flowers and very young fruit. The ripe fruit though, is seldom attacked by birds as it is harvested early, but Leveller, which is left to ripen on the bush, can suffer depredations from blackbirds, thrushes and other birds. Controlling birds is difficult (page 95) and for this reason it is unwise to plant gooseberries near woodlands or in other vulnerable areas.

Site, soil and climate

SITE

As with black currants, the site chosen for gooseberries should be as free as possible from the danger of spring frosts. The requirements for sites for black currants (page 7) equally apply to the gooseberry. The gooseberry flower is slightly more frost hardy than the black currant but it opens earlier, being the first of all fruit crops to come into flower, and so can be frost damaged early in the spring.

Gooseberries can be protected from frost by water sprinkling during periods of low temperature as described for black currants (page 8). However, as the gooseberry and red currant flower over a fairly long period and as these fruits command lower prices, frost protection by fixed sprinkler systems is seldom seen in plantations of these fruits.

In addition to freedom from spring frosts, gooseberries respond to shelter, particularly at blossom time and during the summer when the soft shoot growth is easily damaged by strong winds, making it difficult to produce well-shaped bushes. Where natural shelter is limited, it may be prudent to plant additional windbreaks or even erect temporary artificial screens of plastic netting. If planting new windbreaks there are obvious advantages in getting them well established some years before the land is required for the gooseberry crop (page 6).

CLIMATE

The climate of Britain favours the growth of the gooseberry and this fruit attains a greater perfection here than in almost any other part of the world. An adequate supply of moisture while the fruit is swelling is essential but high temperatures are detrimental. Growth is rapid in spring and early summer during conditions of moderate warmth, moisture and shelter.

The gooseberry succeeds in all the fruit-growing districts of the country, but early production is favoured by a warm spring climate and a southerly aspect.

SOIL

Well drained loamy soils, of good depth and medium texture, with good supplies of organic matter are most suitable. Poor drainage is very detrimental and gooseberry bushes are readily killed by any waterlogging. Sandy soils, very coarse in texture or too shallow, may be excessively dry in early summer and are often low in nutrient elements and will give weak growth and small berries. Heavy clays are generally unsuitable, especially for the varieties Leveller and Careless, though Whinham's Industry is more tolerant of heavy land and indeed may be over-vigorous on deep, well drained soils of high fertility. Shallow soils dry out quickly with a consequent reduction in growth, fruit size and crop weight. Careless makes weak growth on any but suitable soils but is well suited to the lighter and medium silt loams of the Wisbech area. Lancashire Lad thrives on the deep loamy soils of the Lower Greensand in Kent.

The dessert variety Leveller is particularly susceptible to unsatisfactory soil drainage conditions. In Sussex and parts of Kent it does well on soils of the Hastings Beds (particularly on the Pembury series of the Tunbridge Wells sand) which has a fine sandy-loam texture with free drainage.

Gooseberries are tolerant of a fair degree of soil acidity and are fairly resistant to lime-induced chlorosis on calcareous soils, being much less affected than raspberries and strawberries by such conditions.

Varieties

In the last century there were very many varieties of gooseberries listed and a great interest in this crop, particularly in the industrial areas of the Midlands and north, mainly by gardeners seeking to grow prize-sized individual berries. Nowadays the main commercial variety is Careless, a green berry, and only very small areas are grown of other varieties. These include two with red berries, Lancashire Lad and Whinham's Industry; two varieties, Keepsake and May Duke, that may be grown in favoured early areas and whose berries can be picked a few days before Careless, and Howard's Lancer and Whitesmith which make vigorous growth and give larger, later berries than Careless without splitting. The very large, yellow berried variety Leveller requires specialist attention and production, as described above and is restricted to a small area in east Suffolk.

Any variety of gooseberry can be picked hard and under-ripe for culinary or processing use, or it can be left to ripen. When in the ripe, or partially ripe, stage the berries of all varieties are very prone to skin splitting, particularly following sudden spells of wet or warm weather.

Stocks of the lesser used gooseberry varieties are frequently mixed or incorrectly named. There is at present no Certification scheme for gooseberries, but virus diseases, although present, are not so important as in black currants, so most stocks are suitable provided a well grown, disease free bush is obtained.

For pick-your-own enterprises Careless is the most usual variety grown. The fruit can be sold before, or during, the strawberry season and many customers will take gooseberries to help make strawberry jam. It is worth also growing a small area of a red variety such as Whinham's Industry or Lancashire Lad as the fruit is an attraction and, ripening over a long period, can be offered under-ripe for cooking, or left and sold ripe for dessert fruit. It is probably easier to grow these varieties rather than the less vigorous, true dessert variety Leveller. But if the soil is good and the correct pruning can be undertaken this variety will be popular and a small area can well be grown to add to the range of soft fruits offered to the public.

Careless

Bush: moderately vigorous on good soils, weak on poor soils; slightly upright when young, tending to spread later. *Berry:* large, oval, tapering to stalk end; green, pale green to milky-white when ripe; smooth with transparent skin. *Flavour:* good. *Season:* midseason. The most important variety and widely grown for jam-making, canning and for picking green.

Howard's Lancer

Bush: vigorous, upright then spreading. *Berry:* medium, roundish to oval; pale green, slight yellow tinge, skin thin, transparent. *Flavour:* very good. *Season:* midseason. Useful for dessert or for picking green. An old and heavy cropping variety, but stocks are often mixed.

Keepsake (Berry's Early Kent)

Bush: vigorous, rather spreading with slightly pendulous growth. *Berry:* medium to large, oval; green, ripening to whitish-green; slightly hairy. *Flavour:* good. *Season:* midseason. This variety gains size very quickly and can be used for the early green berry trade. Subject to attacks of American gooseberry mildew and one of the most susceptible to frost damage.

Lancashire Lad

Bush: moderately vigorous, upright when young to slightly spreading later. *Berry:* medium large, oblong-oval; deep red; hairy. *Flavour:* fair. *Season:* midseason. Grown either for picking green or for dessert. A popular red variety. Fairly resistant to American gooseberry mildew.

Leveller

Bush: moderately vigorous but rather weak on poor soils; spreading. *Berry:* very large, oblong-oval; yellow-green, almost hairless. *Flavour:* good. *Season:* midseason. This is the most popular dessert variety for commercial growers, but it demands good soil conditions with perfect drainage and specialist culture. Adversely affected by sulphur fungicides.

May Duke

Bush: moderately vigorous, upright. *Berry:* medium to moderately large, roundish oblong; green but ripens to a deep red; smooth, slightly downy. *Flavour:* fair. *Season:* early. This variety is grown mainly in early areas for the green berry trade and is the earliest gooseberry to reach a pickable size.

Whinham's Industry

Bush: vigorous, fairly upright with somewhat arching shoots. *Berry:* medium to large, oval; ripens to a dark red; hairy. *Flavour:* very good. *Season:* midseason. A good all-round variety either for preserving or for the dessert trade. Succeeds better than most varieties on heavy soils but is very susceptible to attacks of American gooseberry mildew. Generally crops more heavily than the other red variety Lancashire Lad.

Whitesmith

Bush: vigorous, upright when young, slightly spreading later. *Berry:* moderately large, oval, pale green with a yellow tinge. *Flavour:* very good. *Season:* midseason. A very good cropping variety and one which may be used either for the green berry or the dessert trade, but stocks are often mixed.

POLLINATION

As with the black currant the gooseberry is self fertile and does not need cross-pollination to ensure a crop. The gooseberry flowers so early in the season that hive bees are seldom seen on the flowers. The pollen grains may be transferred to the stigmas in the flower to ensure fertilisation by wind, as well as by bumble and other bees; also by other insects which are active under the cool conditions of late March and April during which most gooseberries are in flower.

Propagation

FORM OF BUSH

The requirements of a suitable site for gooseberry propagation are the same as for black currants (page 22). Also like black currants gooseberries are propagated mainly by hardwood cuttings (page 22) but they do not always root so freely and often the percentage take is disappointingly low. Gooseberry bushes can be produced by layering the shoots, but this gives a bush without a suitable straight stem or leg, so is not as popular as cuttings.

Unlike the black currant, the gooseberry is generally grown on a single short stem, or leg, some 15 cm (6 in.) in length. The aim in propagation is to ensure a good strong vertical stem without suckers (shoots) arising on it or from the rootstock and this is completely different to the stooled bush of the black currant. Growing gooseberries on a leg was devised to help in hand hoeing for weed control around the spiney bushes. With the advent of herbicides for weed control there is no longer this requirement, but experienced growers still prefer a bush on a leg as being much easier to pick and spray, and the fruit is less likely to get splashed with soil. Bushes grown as stools, or where suckers are allowed to develop from ground level, become over-grown, dense and difficult to pick. However, stooled bushes may give heavier crops on sites where growth is likely to be weak and they tend to survive longer than bushes on a leg. The brittle stem of a bush grown on a leg is vulnerable to mechanical damage, invasion by *Botrytis* and other fungi, followed by death. If stools are grown, it is essential to avoid the growth of perennial weeds amongst them. When bushes are grown on a leg any unwanted suckers that arise below the head of the bush must be removed by pulling off, by hand or with a suckering iron or spade. If cut off they will regrow the next year.

HARDWOOD CUTTINGS

Cuttings should be of one-year old wood made from strong, straight shoots. Old or weak bushes do not yield shoots of suitable strength for cuttings, but these can be readily obtained when young vigorous bushes are pruned. It may be advisable to use a number of bushes specially grown as stools for the annual production of wood suitable for cuttings; all one-year wood being cut hard back every year to ensure an ample growth of new shoots. Excessive vigour in the stools, however, may produce very strong shoots with pithy centres which are not good material for cuttings. Research work has shown best results are obtained when the cuttings are taken and inserted in the early autumn before all the leaves have fallen from the bushes. Cuttings planted during the winter and spring make less vigorous growth and a proportion may be lost altogether through failure to root. Dry weather in spring may result in the loss of many late planted cuttings.

Cuttings should be at least 25 cm (10 in.) and preferably 30 cm (12 in.) long after preparation. The thin, soft wood at the tip of the shoot is removed by cutting back to a well developed bud and the base of the cutting is cut off close below a bud. It is common practice to remove all the buds, with their accompanying spines, except for the four or five buds at the top of the cutting. This ensures that no branches will arise less than 10–20 cm (4–8 in.) from soil level, and that in consequence the final bush will be on a good leg. Occasionally gooseberries are grown on the stool principle with no clear leg and with branches rising from or below ground level as for black currants. When raising such bushes it is not necessary to remove any buds from the cuttings.

Research work has shown that a higher proportion of rooted cuttings can be obtained if the basal buds are left intact. If cuttings so made are intended to be grown ultimately as bushes on legs, any lower shoots will then have to be rubbed off from the yearling bushes when lifted and before planting in their permanent positions. Cuttings should be prepared and immediately planted; they should not be left to dry out. The cuttings can be mechanically planted as for black currants (page 22), or be planted by hand in a slit made with a spade. Whichever method is adopted it is important to get the cutting vertical and to the right depth, as this will help in the production of a good straight leg. If the cutting is inserted too deeply, the leg will be too short and roots and suckers arise too near the head of the bush. If the cutting is inserted too shallowly, not only is it less likely to root, but the long leg will droop under the weight of new growth and give a bush without a straight leg.

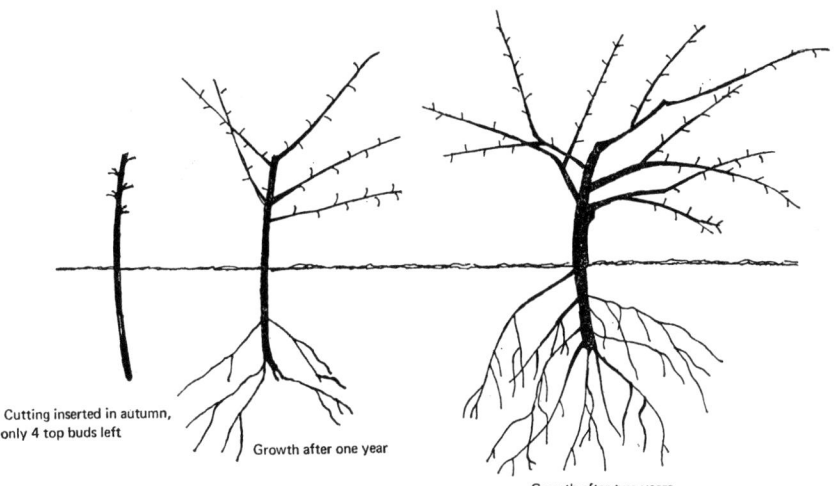

Cutting inserted in autumn, only 4 top buds left

Growth after one year

Growth after two years

Gooseberry bushes are generally planted out in permanent positions when at least two years old. One-year old bushes are seldom strong enough to be set out. There is considerable merit in planting out a well grown three-year old bush, particularly if the plantation is in a rather windy position. The bush head can then be shaped in the shelter of the nursery rows and is less likely to be spoilt by wind damage to shoots when in the plantation.

If the bushes are to be kept for two to three years in nursery rows, the planting distance for the cuttings will have to be wider than for bushes to be left for only one year. But if labour is available excellent bushes can be produced by first rooting the cuttings about 15 cm (6 in.) apart in rows 60–75 cm (2–2½ ft) apart. All rooted cuttings are then lifted at the end of the first season, sucker shoots removed as necessary, and the bushes transplanted back not closer than 30–45 cm (1–1½ ft) apart in rows 1 m (3 ft) apart. The branches are pruned at the end of the next year, and again when lifted as a three-year old bush before planting (page 87). This method enables bushes to be carefully selected and shaped so that only strong well grown three-year old bushes are finally planted out and all have a good leg.

Cuttings should not be taken from bushes or stools infected with American gooseberry mildew and preventative spraying for this disease should be given to the nursery rows (page 92). In addition a routine spray for aphid control should be applied each spring as attacks can deform the shoot growth needed to produce a well shaped bush (page 91). A watch should also be kept to prevent damage from capsids and other insect pests. It is important to ensure that cuttings are taken only from bushes true to type and apparently healthy and vigorous.

Planting

PREPARATION OF SITE

The gooseberry is a long-lived plant and careful preparation of the land will be well repaid. Good drainage is all important and any essential tile draining and ditching should be carried out well in advance. A plough or soil pan if present should be broken up by subsoiling, preferably in dry weather to obtain a good shattering action. Perennial weeds such as common couch or bindweed must be killed, either by using suitable herbicides (page 48) or by means of a bare fallow with suitable cultivations and herbicides.

Finally, the nutrient status of the soil should be checked, corrected where necessary, and the land made into a fit state for planting by early October, exactly as described for black currants (page 30).

PLANTING

All the planting methods detailed for black currants are equally applicable to gooseberries (page 31). The only difference is that since the gooseberry is normally grown on a leg it should not be planted too deeply. All suckers and buds on the roots and lower stem must be pulled off and it may be necessary to remove a few of the upper roots from the main stem before planting. In the few cases where it is desired to grow a stooled bush, the lowest branch should be planted just below soil level. A light cultivation may be necessary to level the soil after planting. Thereafter the plantation should be maintained in a weed free condition with suitable herbicides (page 45).

PLANTING DISTANCES

Although planting on the square is still done, the most usual system is a rect-angular plant, often in a three-row bed system. As with currant plantations, the distance apart of the rows will be dictated by the size of tractor used for spraying and the expected growth of the bushes. Bushes are now grown in herbicide treated soil so that in-row cultivations are not needed.

At present gooseberries are still picked by hand. The black currant straddle harvesters may be adapted to pick this crop in the future, although the fruit is carried rather near to the ground. If successful mechanical harvesting is introduced, adequate space must be left between the rows and the considerations for machine harvesting discussed for black currants will also apply (page 29).

Gooseberries are usually grown in rows 2·4–2·7 m (8–9 ft) apart. The bushes are kept distinct about 90–120 cm (3–4 ft) apart and are not usually allowed to form a continuous hedge as with black currants. These distances will be suitable for Careless, May Duke and similar varieties, but Lancashire Lad, Whitesmith and Keepsake may need slightly wider spacing.

If planting three-row beds, the centre row is put at half the row distance and the bushes may be staggered to come opposite the space between the bushes in the two outside rows, thus aiding access by pickers. Spraying, as for currants, of the centre row is done from both sides. If the bushes become too crowded in after years, the centre row can be removed. The number of bushes required to plant areas at various spacings is given on page 27. Whatever planting systems are adopted, too close a spacing between bushes will lead to interlacing of the spiny branches, which can make picking very difficult.

Because of the difficulty in rooting gooseberry cuttings and the need to plant a well shaped bush, plantations are not established direct from cuttings as in the case of black currants.

When planting gooseberries as undercrops to plums or apples, the bushes should be spaced a little wider and fitted in as convenient to the main top fruits, which must dictate the row and tree position. The gooseberry bushes might be spaced as a single row within the top fruit row. More often an additional row of bushes is planted on each side of the top fruit row, about 1·5 m (5 ft) away. Plenty of space should be left around each tree. This method permits the tractor to travel up the alleyways from which both top and soft fruit can be reached for spraying and other operations. By leaving some room around each top fruit tree the danger from spray drip is minimised, if not eliminated. As the trees become established the bushes should gradually be removed, unless it is planned to keep them as a permanent undercrop.

Manuring

The principles of manuring and the main fertilisers providing plant nutrients are described on page 33, but sufficient information is not yet available for leaf ash analysis to be used as a guide to the economic manuring of gooseberries, as in the case of black currants.

Gooseberries have a high requirement for potassium and insufficient uptake can result in poor growth, with the leaves showing a marginal grey scorch or necrosis. Nitrogen is needed, but bushes over-supplied with this element, particularly those on fertile, moist soils, may produce excessive soft shoot growth which is very liable to be smashed by winds in May and June before the wood has lignified. It is considered that the gooseberry has only a limited requirement for phosphorus but any deficiency, as shown by pre-planting soil analysis, should be made good at the time of pre-planting soil management.

The gooseberry has a fairly high requirement for magnesium but this element must be kept in balance with potassium (page 38). The lime or calcium requirement of gooseberries is lower than that for black currants and bushes will grow and crop well on soils with a pH of 6·0–6·5.

The amount of fertiliser nitrogen required for gooseberries, as with black currants, depends on the amount of summer rainfall or irrigation. Where the summer rainfall (April–September) is below 350 mm (14 in.) 100 kg/ha (80 units/ac) N should be applied per annum. If the summer rainfall is over 350 mm (14 in.) 50 kg/ha (40 units/ac) N should be applied per annum. These rates can be reduced if bulky manures are used or if the soil is very fertile, also if the growth is too vigorous.

PHOSPHORUS

This element is not so important for gooseberries as for black currants, but it is needed for root production, particularly at the time of planting. In order to maintain fertile soil the annual rates recommended for black currants (page 36) should be provided for gooseberries.

POTASSIUM

As previously emphasised this element is very important for gooseberries and so the annual rates recommended for black currants should be applied to gooseberries (page 37). In the past it was suggested that potassium should be given to gooseberries only as sulphate and not in the chloride (muriate) form, because of the danger of necrosis. However, for many years growers have been using compound fertilisers to apply the necessary amounts of NPK to bush fruits and in most of these fertilisers the potassium is in the less expensive chloride form and there have been no reports of damage. It would, however, be wise to use the sulphate form if the pre-planting soil indices for this element (page 37) were below 2, indicating that considerable amounts of potassium were needed both before and after planting.

MAGNESIUM

Reference has already been made to the fact that the gooseberry has a fairly high requirement for magnesium. The need for a balance between potassium and magnesium in the soil and the fertiliser practice to be adopted is discussed on page 38. The same annual amounts of magnesium as for black currants are recommended for gooseberries (page 37).

If the soil is acid, so that lime is required before planting, magnesian limestone should preferably be used, as this will help to increase the amount of available magnesium (page 36) in the soil.

As with any fruit crop routine soil sampling and analysis (page 35) will enable an assessment to be made of the index (nutrient status) of the soil, so that the annual fertiliser rates can be adjusted accordingly.

As the gooseberry comes into leaf early in the year it is important to apply fertilisers—if using a manure spinner—before leaf break or the fertiliser may lodge on the leaves and cause damage. If applying fertilisers by hand after leaf break avoid wet and windy days.

Straw mulching the rows will be beneficial for gooseberries as for black currants, but care must be taken to apply additional nitrogen to aid the decomposition of the straw and so avoid soil depletion of nitrogen (page 35). There is some experimental evidence to suggest that heavy mulches of farmyard manure, particularly under wet winter conditions, may be detrimental to the growth and cropping of gooseberries. In general this material would be better used to improve the soil by incorporation before planting.

Pruning

The aim in pruning gooseberries is to form a vase or cup shaped arrangement of the main branches on which the numerous spurs and short laterals (one-year old shoots) arise which carry the crop. The main framework is permanent, although as the bush ages it will be necessary to remove broken, dead, or badly placed branches. It is thus important in the early years to ensure a well spaced main branch system by selecting the best placed shoots and shortening them so that they are stiffened and grow in the right position. Some gooseberries, particularly those for pick your own, are now being trained on a low wire trellis as described on page 101 for red currants. This avoids wind damage and gives earlier crops.

Pruning is best left until the late winter, for if there has been bud removal by bullfinches this will have to be taken into account when deciding where to make the pruning cuts. All cuts should be made carefully with sharp secateurs just above a bud that points in the direction in which it is hoped the shoot will grow the following year.

PRUNING THE YOUNG BUSH

All varieties are treated in the same manner for the first three years. The cutting should have made three to four new shoots from the buds left intact when it was taken. Only the best placed shoots are cut back to 15 cm (6 in.). Any that are weak or badly placed are completely removed. At the end of the next year's growth there should be a well spaced whorl of six to twelve shoots. Any that are badly positioned are cut away and the spaced shoots needed for the branches are again each cut back to leave about 15 cm (6 in.) of new growth.

At this stage if the bush is strong it may be planted in its permanent quarters, or if weak, left to grow one more year in the nursery in order to achieve a well rooted bush with a good thick leg and a well positioned head of six to twelve branches.

Leader pruning, which allows new extension growth on the end of each branch of 15–30 cm (6–12 in.), more if strong, less if weak, is continued until the bush reaches its final size. During these years there will be a need to select shoots to form additional branches and to leader prune or tip these as necessary. All other strong shoots that are produced in the bush and which are not required to form a branch are regarded as side shoots, or laterals, and pruned as described in the next section.

PRUNING THE ESTABLISHED BUSH

All varieties, except Leveller, are treated in the same manner. The gooseberry carries its fruit along the new growth produced the previous year and also on spurs on the older wood. The leading shoot of each branch is tipped annually as already described but once the bush has reached a reasonable size no further leader pruning is required because the leader will grow less strongly and in time produce only a short growth each year that does not require tipping and which will carry some fruit.

The laterals or side shoots, are of three kinds:

1. Short stubby spurs that, growing very slowly each year, produce plenty of fruit buds and need no pruning.
2. Weak one-year shoots about 10–20 cm (4–8 in.) long that generally need no pruning.
3. Stronger shoots that can be left full length to crop, if room permits, or cut right out, back to the branch from which they arise, if growth of the bush is too thick. Crossing, drooping or generally badly placed shoots within the bush should also be cut right out.

As the bush ages the centre may become overcrowded with shoots and so difficult to pick. It will then be necessary to cut right out one or two major branches together with all their laterals. This can best be done with hand loppers.

At all times any weak drooping shoots, or branches, should be removed to keep the fruiting branches off the soil. Any strong down-growing new shoots should also be removed, preferably by snapping and twisting them out from the base with a well gloved hand, rather than by cutting off. If cut-off, new shoots may grow the following year from buds left at the base of the cut-off shoot.

Suckers arising from the rootstock, or the leg, will need removal, if possible cutting or pulling off at the base so as to avoid regrowth.

PRUNING LARGE COMMERCIAL PLANTATIONS

With increasing labour costs growers have tended to reduce the amount of detailed pruning as described above, and by providing good growing conditions ensured plenty of new, strong, but balanced shoot growth. Very little leader tipping is necessary after the first three to five years and the bush centre is kept open by pulling and twisting out unwanted shoots, also those growing from below the head. This work can be readily done by a worker, with well gloved hands, as the new shoots are weak at the point of attachment to the older wood. When necessary larger branches should be cut out with loppers. If the workers that do the pruning also help with the picking they will appreciate the need for adequate access both to the centre of the bush and to each branch.

Old branches that have stopped growing and are tending to give poor crops of small berries can be replaced by training in new vigorous shoots from time to time, provided the bush itself is still growing vigorously.

Keepsake and Careless are usually pruned rather severely, but the stronger growing Whinham's Industry and Howard's Lancer more lightly. The fruits of certain varieties which decline appreciably in vigour, such as Whitesmith, rapidly lose size with age if pruned too lightly.

Excessive pruning results in too many young vigorous shoots which are likely to be unfruitful and liable to attacks of disease and to blow out in winds. The severity of pruning should be governed by the vigour of the bush and the response made to the pruning carried out the previous winter.

PRUNING OF LEVELLER

Leveller has to produce really large dessert fruits for profitable culture. Pruning therefore has to be severe to limit the amount of fruit produced by each bush and so increase individual berry size. Pruning in the early years is exactly as described above, the leaders being tipped to stiffen and space the branches and new branches being formed as the bush grows. This procedure is continued throughout the life of the bush.

All the laterals arising on the branches are treated more drastically than for other varieties. Natural spurs are left uncut, but all the other laterals, even if fairly short, are cut back to 3 cm (1 in.) of new wood each year, so that all the fruit is carried on short spurs situated on the main branches.

REMOVAL OF PRUNINGS

Because of their spines the prunings are generally removed from the plantation, using a buck rake to push them out, rather than leaving them in the alleyways and pulverising them as can be done with black and red currants. Gooseberry prunings left about will seriously hinder the pickers who will often have to kneel or sit on the ground to harvest the crop.

Control of weeds

The gooseberry is grown under a system of non-cultivation and the principles of the use of herbicides and the types are fully discussed on pages 44–52. All of the herbicides used for black currants can also be used for gooseberries, though chlorthiamid and dichlobenil may cause leaf margin chlorosis in the season of treatment or later; but this does not usually indicate an adverse effect on growth or cropping. Do not apply chlorthiamid to gooseberries growing on light soil. The dormant buds of gooseberries are resistant to paraquat so that, unlike black currants, gooseberries can be safely sprayed overall with this herbicide provided the buds are dormant and it is needed for the control of annual weeds or creeping buttercup. For gooseberry cuttings weed control is the same as for black currant cuttings described on page 49.

Harvesting

The first gooseberries are picked while they are hard green and as soon as they have achieved a size of about 14 mm ($\frac{1}{2}$ in.) in diameter. The berries are still growing rapidly at this period in late May, and the loss of crop through early picking has to be balanced against the higher prices secured for these earliest berries. The first gooseberries are usually sold loose in 5 kg (10 lb) trays or baskets although they may be packed in 250 or 500 g ($\frac{1}{2}$ or 1 lb) punnets in trays. The bushes can be picked over several times removing the largest berries; this encourages more rapid growth of the remaining berries. Mechanical regrading of berries is not usually required; the pickers are shown the minimum size of berry to gather.

The main commercial areas are cleared at one picking when the fruit has attained maximum weight but the berries are still firm. The organisation of pickers is described on page 64. It is important not to misjudge the time of harvest, for if weather conditions change and harvesting is delayed the crop may become over-ripe, the berries split and so become difficult to market.

The berries are usually picked piecework, payment being made for each full bucket. Gooseberries can be picked much more quickly than black currants and, depending on crop weight and berry size, about 100 picking hours is needed per tonne (ton) of berries. A good worker can pick about 11 kg (25 lb) per hour. The pickers bring the buckets to a suitable tally point where the fruit is tipped into sacks. Pickers must be instructed to keep leaves and broken shoots out of the buckets and to pull the berries off with the minimum of damage to the bushes. A small fan blower situated above a simple chute, down which the berries are tipped from the buckets into the sacks, can aid leaf removal. The sacks are then weighed, usually to a constant weight of 25 kg (56 lb), and stacked on pallets to await collection. Pallets should be stood in the shade as soon as possible and kept dry, for although hard, the berries will soon start heating and sweating if left too long in stacks under unsuitable conditions.

In 1976 a modified black currant straddle harvester was used to harvest a Careless plantation. One shaker drum was used in a horizontal plane over the bushes and the conveyors delivered the fruit to a bulk bin. Preliminary observations showed that sixty-five per cent of the fruit was harvested, the rest being left on the bushes or ground. Further experience and, if necessary, adjustment to the machine should improve the percentage harvested.

The dessert variety Leveller needs more careful harvesting. Careful pruning should ensure that all the berries are of suitable large size. Any small berries can be placed in a separate container as the crop is picked, or all the crop can be cleared and the fruit sized, either with a small mechanical sizer or by eye, and the dessert fruit placed into 500 g (1 lb) punnets. The smaller berries are generally sold in 5 kg (10 lb) trays or baskets. If the berries ripen unevenly this variety may need picking over two or three times. It is important to harvest the berries immediately they are just ripe for warm, moist weather will encourage the ripest berries to split either on the bush or during marketing. Over-ripe and fully ripe berries should not be dispatched to distant markets but sold locally.

Control of pests and diseases

The spray machinery described for black currants is suitable for gooseberries which, because of their stiffer branches and less dense growth, are an easier spray target. Gooseberries have few serious pests and diseases but in most seasons, and particularly if it has been noted the previous year, routine protective spraying will be necessary for the control of American gooseberry mildew, which is the most serious disease. Amongst pests it is wise to take annual precautions for aphids and for sawfly, which can do remarkable damage by defoliating bushes very rapidly if not checked in time. Sawflies, however, do not occur in every plantation. Adequate inspection of all bushes should be made to detect the early symptoms of damage. Where the pest has not previously occurred control measures need only be applied directly damage is seen (see below).

Pests

Aphids (Leaflet 176)

Several species of aphid occur on gooseberries, including the gooseberry aphid (*Aphis grossulariae*) and the currant-lettuce aphid (*Nasonovia ribisnigri*); the latter species rarely causes serious damage. The gooseberry aphid may cause severe distortion of young shoots and leaves from May to July. Control measures are similar to those given for aphids on currants (page 54).

Capsids (Leaflet 154)

The common green capsid (*Lygocoris pabulinus*) can be a serious pest of gooseberries as well as black currants (page 56).

Mites (Leaflet 305)

The gooseberry red spider mite (*Bryobia ribis*) was formerly a serious pest but is now uncommon in gooseberry plantations. It overwinters in the egg stage and infestations develop in the spring or early summer; there is only one generation in a season.

Organophosphorous pesticides applied at first open flower against aphids are effective against this pest and this probably accounts for its decline in importance. Should heavy infestations still develop a spray should be applied as soon as mites are observed on the opening buds. Suitable materials include demeton-S-methyl, dimethoate, formothion and oxydemeton-methyl.

Moths

The currant clearwing moth (*Synanthedon salmachus*) can be a problem on gooseberry plantations in some areas (page 56). The magpie moth (*Abraxas grossulariata*) was formerly an important pest but is now rarely seen in commercial plantations (Leaflet 30).

Sawflies

Three species of sawfly occur on gooseberry, but the commonest and most important is the gooseberry sawfly (*Nematus ribesii*). Adults appear in April and May depositing eggs on the underside of leaves, especially low down in the centre of bushes. The caterpillars feed avidly on the leaves and later spread throughout the bush, eventually pupating in the soil. Three overlapping generations occur in a season. Good control is achieved by spraying with azinphos-methyl plus demeton-S-methyl sulphone, chlorpyrifos, derris, fenitrothion or malathion as soon as the caterpillars are seen, this is usually in mid-May. A second spray two or three weeks later may be necessary.

Diseases

American Gooseberry Mildew (Leaflet 273)

This disease, caused by *Sphaerotheca mors uvae*, affects leaves, stems and fruits. First signs are seen as white powdery patches on the young leaves soon after they unfold in the spring. These powdery areas produce numerous spores and, under favourable conditions, the disease soon spreads to developing fruits and extension shoots where further powdery areas are produced. As the season progresses these areas become matted and change colour to fawn and finally chocolate-brown.

Fruits infected early in their development are retarded in growth and become enveloped in the white or brown fungal mycelium; later infections produce unsightly areas which, although superficial, render the fruit unmarketable. Infection can also cause young shoots to be stunted and deformed, with consequent reduction in crop the following year.

Varieties vary in their susceptibility to the disease. Keepsake and Whinham's Industry are among the more susceptible, while Lancashire Lad is relatively resistant. American gooseberry mildew can also attack black and red currants but on the latter it is rarely severe.

Soft sappy growth under conditions of high humidity is most susceptible. Balanced manuring, combined with judicious pruning to keep the bushes relatively open, can therefore do much to lessen the risk of infection. Where these measures fail, protective spraying with a fungicide is advisable. Several fungicides are effective if applied as soon as the fruit is set and again three weeks later; in severely attacked plantations an earlier spray may be necessary just before flowering. The recommended spray programme will protect the berries but may not give adequate control of shoot infections in severely affected plantations. In these instances, removal of infected shoot tips in late August or September will be beneficial.

Benomyl, carbendazim, dinocap, quinomethionate or thiophanate-methyl can be used successfully to control mildew. Dinocap is generally safe on all varieties, although occasional instances of damage to Leveller have been recorded; an additional wetter is necessary with some formulations of this material if results are to be satisfactory. Sulphur dust, lime-sulphur and various proprietary sulphur preparations will also give control, but several varieties, including the widely grown Leveller and Careless, are sulphur-shy and may be damaged by these materials. If the crop is intended for processing, the processors should be consulted before applying any spray.

There is some evidence that mildew tolerance to benzimidazole type fungicides (benomyl, carbendazim and thiophanate-methyl) can occur, which limits the effectiveness of this group of chemicals. Growers should consult with their spray chemical firms' representatives, or ADAS, about spray programmes to avoid the continuous use of these particular chemicals.

European Gooseberry Mildew (Leaflet 273)

Microsphaera grossulariae, a fungus related to that causing American gooseberry mildew, also attacks gooseberries to produce a delicate white mould growth usually restricted to the upper leaf surface. This disease is not very common and rarely causes significant damage. It is controlled by the measures recommended above.

Die-Back

Gooseberry plantations are commonly affected by die-back. A number of fungi may be associated with the condition but in the most severe cases the grey mould fungus, *Botrytis cinerea*, is often implicated.

Symptoms may appear at any time during the growing season. Individual shoots and branches at any position on the bush may be attacked and the first indication is the collapse and death of the leaves. The bush may resist further attack but frequently infection extends from one branch to another and the whole bush is progressively killed. Infection may start on the main stem near soil level and if the stem becomes girdled the whole bush dies. It may take a year or more for this to occur and a half dead bush is a common sight. The bark of affected branches splits and peels to expose the underlying tissues, in warm moist weather characteristic grey fluffy tufts of *Botrytis* can sometimes be seen growing from them. When sporing ceases, black resting bodies or sclerotia may be formed embedded in the bark. Sometimes the berries are attacked, small rusty-brown patches appearing which spread through the maturing fruit to cause a soft rot. In exceptionally favourable conditions, the fungus may spore all over the infected fruit. The dead flower remaining on the tip of the fruit is sometimes a point of entry for the fungus to cause a fruit rot.

Although die-back may be found in a variety of circumstances, it is most severe on bushes growing in unsatisfactory conditions. Soft sappy growth resulting from excessive nitrogenous manuring is very liable to attack, as are bushes deficient in potash. Poor drainage weakens the bushes and increases liability to infection. The fungus more readily gains entry through wounds than by directly infecting healthy tissue, and care should be taken to avoid unnecessary damage during cultivations, picking and pruning. All dead and dying branches and bushes should be removed from the plantation and burnt. Fruit rot can be controlled by applications of benomyl, carbendazim, dichlofluanid or thiophanate-methyl. There is some evidence that die-back tolerance to benzimidazole type fungicides (benomyl, carbendazim and thiophanate methyl) can occur, which limits the effectiveness of this group of chemicals. Growers should consult with their spray chemical firms' representatives, or ADAS, about spray programmes to avoid the continuous use of these particular chemicals.

Coral Spot (Leaflet 23)

Coral spot, caused by *Nectria cinnabarina*, also occurs on gooseberries in association with die-back but it is less common than *Botrytis*. The presence of small, salmon-coloured, spore-producing pustules on the dead and dying branches is characteristic and, in spring, some of these may later turn dark red to produce a second type of spore. Measures taken against *Botrytis* die-back also help to control this disease.

Honey Fungus (Leaflet 500)

The honey fungus, *Armillaria mellea*, can cause serious loss in plantations sited on land previously cleared of scrub, woodland or old orchards. Affected bushes are conspicuous by their sickly appearance, later followed by complete collapse and death. Usually groups of bushes are affected, but individual bushes may also be attacked. Black rhizomorphs (bootlaces) of the fungus can sometimes be found on the surface of roots of dead and dying bushes and under the bark fan-shaped masses of white mycelium, with a distinct mushroom odour. The rhizomorphs spread the disease from one bush to another. Infected bushes should be dug up and burnt, leaving in the ground as few root fragments as possible.

Cluster Cup Rust

The rust, *Puccinia pringsheimiana*, appears in early summer as smooth red or orange swellings on the leaves and fruits and occasionally on the stems. The smooth surface of these swellings later becomes pitted owing to the development of numerous cups, each the size of a pin head, with a pale reflexed fringed margin. These fructifications produce an abundance of yellow spores which infect not the gooseberry but various species of sedge. On this secondary host two types of spore occur, uredospores which spread the disease from one sedge plant to another and teleutospores which are the means of producing yet another kind of spore, capable of infecting the gooseberry. As might be expected, the disease is worst in the neighbourhood of wet or marsh areas where sedges are growing. In some instances, control of the sedge is practicable, but otherwise such areas should be avoided if possible. The disease is seldom serious enough to warrant special control measures.

Another rust disease, caused by *Cronartium ribicola*, very occasionally infects gooseberries. It is much more common on black currants.

Leaf Spot

Leaf spot, *Pseudopeziza ribis*, similar to that on black currant, frequently affects gooseberry but is usually less severe. In some instances, however, it causes premature defoliation and weakens the bush. Control measures are similar to those recommended for black currant on page 58.

Vein-Banding Virus

Almost all gooseberries in commercial production are infected with vein-banding virus, which causes translucent, pale yellow banding along the main veins. In the first leaves to expand in spring all of the veins may be affected, whereas in leaves on extension growth only single veins or short lengths of the main veins may be affected. Leaves with vein-banding are often yellowish and asymmetrical. Symptoms of the virus can be confused with those of aphid feeding, especially by *Hyperomyzus lactucae*, which causes diffuse yellow vein-banding usually accompanied by interveinal mottling. Comparison of healthy and infected seedlings suggests that infection reduces vigour. In the past, growers may have inadvertently eliminated virulent strains of vein-banding by propagating from vigorous bushes free from abnormalities; rigorous selection on these lines should be continued. Some varieties, such as Leveller and Lancashire Lad, seem more sensitive to infection than others (Careless). Black currants are sometimes infected with gooseberry vein-banding virus and red currants commonly suffer from what seems to be an identical graft-transmissible disease.

Experiments have recently shown that elimination of vein-banding from Careless by meristem culture has appreciably increased yield and fruit size. Virus-free material of this and other varieties should eventually become commercially available. It will then be important to delay reinfection by isolation from infected crops (gooseberries and currants) and by strict control of the aphids which spread vein-banding. These include *Aphis grossulariae, A. schneideri, Hyperomyzus lactucae* and *Nasonovia ribisnigri*. Spraying gooseberries to control aphids is already worthwhile because of the direct damage they cause (page 91).

BIRD AND ANIMAL DAMAGE

The dormant winter buds of gooseberries and red currants and sometimes black currants are frequently seriously damaged by bullfinches. Gooseberries are particularly susceptible and the damage not only seriously reduces the following crop, but can spoil the whole shape and growth of the bush, since new growth and hence shoots do not arise from the nodes. Damage may take place in December or more usually after Christmas. Sparrows may attack the flowers of currants and gooseberries but their depredations are more erratic. Bullfinches can be controlled by trapping and shooting in areas where local by-laws permit.

Small areas of gooseberries or red currants, where damage is regularly expected, can best be protected by training a proprietary protective nylon webbing over the tops of the bushes in the autumn. The mesh is teased out by hand over the rows but as it is not removed in the spring it can be a nuisance during picking. Some growers use instead five or six spindles of black cotton on a pole with threads about 15 cm (6 in.) apart. The pole is drawn over the row, thus teasing out the threads over the bushes. Both methods sometimes give little or no protection.

There is no reliable spray material that will give sufficient protection. Proprietary sprays based on thiram will deter attacks but as they are washed off by rain will need respraying. Similarly tar oil winter washes, applied when the buds are dormant, may give some initial protection during the winter.

If bullfinches have proved serious every effort should be made by shooting and trapping from August onwards, where this is legally possible, to reduce the population before damage is done. Bush fruits are not very susceptible to predations caused by rabbits, hares, voles or deer.

Economics

The economics of gooseberry production can be calculated exactly as for black currants (see page 70). Bushes are more expensive than black currants and take much longer to come into full crop. It may well be five to seven years before the plantation is in full crop and the ground covered with fruiting bushes. However gooseberry bushes can remain productive for very many years; plantations well over 20 years old are not uncommon. The higher establishment costs could thus be written off over a 15 year period.

The annual variable costs can be calculated as for black currants and it will be noted that the spray programme could be less expensive than that for black currants. Adequate allowance for suitable herbicides must be included, as weedy gooseberry plantations are particularly unthrifty and the less dense foliage, compared to black currants, may allow weeds to become established more easily.

Pruning can be rather more laborious than for black currants. Allow 60–250 hours/ha (25–100 hours/ac) increasing during the period from planting to about seven years of age. Once the bushes are established, some growers prune carefully using about 250 hours/ha (100 hours/ac), others do very little.

Hand harvesting gooseberries is much faster than black and red currants. Provided the crop is heavy and the bushes well shaped experienced pickers should pick about 11 kg/hour (25 lb/hour) or even more.

In 1976 the first attempts to harvest gooseberries mechanically were made with a modified black currant straddle harvester, with reasonably promising results of an output of about 3 tonne/hr. Since then a few purpose built, straddle harvesters have been introduced with horizontal shaker arms. They are performing well with a high percentage catch of berries when used in suitable plantations.

The handling, weighing and stacking costs for gooseberries for processing are smaller than those for black currants, since gooseberries are generally sold in sacks provided by the processor. If sold on the wholesale market costings are as described for black currants (see page 75).

Red Currants

Growth and crop

The red currant is the least important of the bush fruits. The demand for the fruit is very limited, only a small quantity being required each year for processing mainly for red currant jelly. There is no other major processing outlet; the fruit is not used at present for commercial juice production, unlike the black currant, nor for canning, and only small quantities are needed on the wholesale fresh fruit markets.

In 1939 920 ha (2300 ac) were grown but by 1965 the area had decreased to 230 ha (570 ac) and after this date separate statistics were not kept. Both red and white currants are now included in a general category of 'other small fruits'. These fruits are grown on holdings situated in the main fruit growing areas where other soft fruits are also produced and convenient to processing outlets. Red and white currants, like gooseberries, are long-lived bushes. Recently there has been some limited new planting in the Midlands, Kent and East Anglia using some of the newer varieties. It would be very unwise to plant red currants without first ensuring that there was a reliable outlet for the anticipated crops.

There is a small demand for red currants on farms specialising in pick-your-own soft fruits. Most of the fruit will be used for jelly, juice and wine making and only a limited area should be planted until the likely demand can be assessed. The value of red currant juice to aid strawberry jam making can be advertised on these holdings and should help to increase sales; for this purpose one of the early varieties should be grown.

The red currant is a fairly easy fruit to produce, except for problems from wind and birds, and does not require so much spraying as black currants and gooseberries to keep the crop free from pests and diseases.

Yields can be regular provided there is no frost damage and well grown plantations should yield upwards of 10 t/ha (4 ton/ac), although the average yield is only half this figure. In years past, when demand was poor, not all the crop was harvested. Fruit left on the bushes depresses the succeeding crop.

Site, soil, climate

The red currant flowers as early as the gooseberry and in a precocious spring some flowers may be open in the last week in March—and a little before the black currant, but the flowers are more resistant to frost damage. Even so the flowers can be killed by severe spring frosts, so that the requirements for a suitable site for black currants (page 8) apply equally to red currants.

Red currant shoots grow vigorously in the spring and, like gooseberries, may be smashed by wind before they are lignified. Well sheltered sites are required or artificial, or growing, windbreaks must be provided or the bushes trained on a wire trellis. Like the gooseberry the red currant may be grown as an undercrop to top fruit, chiefly plums and apples. This was frequent practice in the past but is not now recommended.

The red currant flowers and fruits on both old wood (spurs) and the young wood formed the previous year, so it is more akin to the gooseberry than the black currant. Very fertile soils are not required and may lead to too much vigorous growth and bushes difficult to prune and train. But if the soil is very light, and particularly if the soil water supply in summer is likely to be deficient, the berries will be very small unless irrigation can be provided. Red currants can be grown on a wide range of soils but crop best on well drained, medium loams: they are reasonably tolerant of both acid and alkaline soils.

Red currants can be seriously damaged by bullfinches attacking the dormant buds during the winter; they are frequently attacked before any other fruits. Care must be taken not to establish a plantation adjacent to woodland, or hawthorn hedges, which encourages this pest and it will be necessary to keep the bullfinch population down to avoid unreasonable damage (page 95).

Varieties

There are many varieties of red currants, some that have been grown since the last century and are still being propagated. Recently there has been some interest in Europe in this fruit for juice and some new varieties have been introduced that would be well worth testing more widely under U.K. conditions.

There is a fair range in the appearance, type of growth, leaves, flowers, strigs and berries of red currants and so varieties are readily identified; stocks true to type should therefore be obtainable. Red currants are self-fertile and pollinated by early flying insects and wind.

For many years the old variety Versailles was widely grown but this was replaced by Laxton's No. 1. American varieties were imported and tested during the 1950s and Red Lake looked promising, with very long strigs of good sized berries. In these trials Earliest of Fourlands also cropped heavily but the berries were small. In more recent trials the Dutch variety Rondom has given good yields and growers are also interested in the Dutch variety Jonkheer van Tets because of its good crop and earlier ripening season.

In all these variety trials the bushes were pruned to give traditional spur pruned, vase shaped bushes as described below. Some limited trials and growers' experiences suggest that this hard pruning may be reducing the crop and, particularly where berry size is not important, much lighter pruning may be beneficial, provided the vigour of the bush can be maintained. It will be necessary to test selected varieties under this different type of bush management.

Red currants have not yet been mechanically harvested as the area grown is so small. It is considered that if the bushes were trained more as stools it would be possible to harvest them with existing black currant mobile harvesters. For this purpose varieties with shorter strigs that ripen evenly, and with vigorous, upright growth of non-brittle wood should be selected for trial. If there was a demand for red currants for juicing this method of management might well be considered.

Earliest of Fourlands

Bush: vigorous, upright. *Truss:* moderately long, much bunched. *Berry:* medium to moderately large, bright red, seed small. *Quality:* good. *Season:* early ripening, just before Laxton's No. 1. Generally a heavy cropper.

Jonkheer van Tets

Raised in the Netherlands. *Bush:* upright, vigorous and liable to wind damage. *Flowering:* early. *Truss:* long to very long. *Berry:* large or variable size, average quality. *Season:* early, ten days before Laxton's No. 1. A heavy cropper, but the berries are likely to split in wet weather. A useful early variety with attractive fruit.

Laxton's No. 1

Bush: vigorous, upright to slightly spreading. *Truss:* moderately long, much bunched. *Berry:* medium, bright red, seed small. *Quality:* good. *Season:* early. A very heavy cropper.

Laxton's Perfection

Bush: vigorous, upright growth very stout. *Truss:* long to very long fairly free. *Berry:* very large, held laxly; good colour. *Quality:* good. *Season:* midseason. Quite a good cropper but has gone out of favour owing to the production of a percentage of blind buds and to a substantial number of the shoots blowing out. Of interest to amateur gardeners because of the very large berries.

Minnesota No. 71

Raised in U.S.A. at the University of Minnesota this variety has cropped well in a recent trial. *Bush:* upright, vigorous with shoots that are not so liable to wind damage as most varieties. *Flowering:* mid-late. *Truss:* medium. *Berry:* large, attractive, good quality. *Season:* late.

Red Lake

Another variety introduced by the University of Minnesota. *Bush:* moderately vigorous, upright. *Truss:* very long, free, with basal berry not interfering with picking. *Berry:* very large, of even size; bright red, juicy; moderately large. *Quality:* good. *Season:* midseason to late. Apparently shows no tendency to blow out. New variety, very free cropping, producing a remarkably long truss of evenly shaped attractive fruit.

Rondom

Raised in the Netherlands. *Bush:* vigorous and upright with shoots resistant to wind damage. *Flowering:* midseason to late. *Truss:* medium with long handle, the berries tightly packed. *Berry:* medium, variable, thick skinned. *Season:* late, hangs well. Has cropped more heavily than Laxton's No. 1 in a recent trial but stocks are mixed and it is important to obtain a true, heavy cropping clone.

Stanza

Recently raised and introduced in the Netherlands. *Bush:* vigorous with stout upright shoots. *Flowering:* late. *Truss:* short. *Berry:* small-medium, dark red, rather acid, of only moderate flavour. *Season:* midseason to late. Cropping is said to be good and the variety is of interest for good colour of juice.

Wilson's Long Bunch

Bush: moderately vigorous, semi-erect. *Truss:* medium length, rather bunched. *Berry:* medium; bright red. *Quality:* good. *Season:* late. A very free cropping variety, quite widely grown and valuable on account of its lateness. There are several different clones distributed under the same name but the true stock is distinct and can be distinguished by its late ripening habit.

Propagation

The red currant is generally grown on a leg and the method of propagation by hardwood cuttings is exactly as described for gooseberries (page 82). Red currants root more readily than gooseberries and early autumn planting is not so important. The rooted cuttings grow more strongly than gooseberries and it is usual to plant a one- or two-year old bush. Any buds or suckers arising from the rootstock or below the head of the bush should be removed when the bush is set out in the permanent position. Some growers, however, are producing red currant bushes on the stool system as this gives earlier and heavier cropping, but to help the strong young shoots carry the crop they may have to be supported by wirework. Cuttings for stools should have all the buds left intact when being prepared, whilst for bushes to be grown on a leg only the top four buds are left (page 83).

Planting

Thorough pre-planting soil cultivation and preparation is necessary, perennial weeds should be eliminated, soil samples taken and the nutrient status of the soil corrected as necessary (page 12).

Red currants grown as a bush on a leg are generally planted exactly as described for gooseberries (page 84). If the bushes are to be grown as stools, lightly pruned and with the shoots supported by wirework, the row distance will be dictated by the machinery available. If narrow tractors are to be used and the crop is to be hand picked these rows need to be 2·4–2·7 m (8–9 ft) apart with the bushes about 1·2 m (4 ft) apart, but the rows will have to be wider if standard tractors or mechanical straddle harvesters are to be used.

Manuring

Red currants have similar nutrient requirements to gooseberries, with a high requirement for potash (page 37). Care should be taken not to over-manure and so encourage the production of soft new shoots which are both difficult to train and prune, and which may break out from the older branches in windy conditions.

The annual fertiliser programme recommended depends on the rainfall. Where summer rainfall is less than 350 mm (14 in.) 100 kg/ha (80 units/ac) nitrogen should be applied and 50 kg/ha (40 units/ac) nitrogen where the summer rainfall is over 350 mm (14 in.). Phosphate, potassium and magnesium should be applied as for gooseberries (page 85–87).

Pruning and training

The fruit of red currants is mainly carried by buds which arise around the base of the shoots that grew the previous year. Also in the terminal buds on shoots which can arise from the spur system and be only a few mm long, or from dards 10–15 cm (4–6 in.) long.

If the bush is to be grown on a leg with a vase shaped top of radiating branches the leader pruning necessary is similar to that described for gooseberries (page 87). Red currants produce stronger one-year old shoots than gooseberries and it is easier to shape and build up the open centre arrangement of branches by selective pruning of the leader shoots each winter. Any very strong vertically growing shoots are best cut right away from the main framework or they may take over from the more angled, and hence weaker, branches.

On the radiating branches will arise natural short fruit bud producing spurs, and also other shoots of varying lengths. Short shoots or dards, less than about 15 cm (6 in.) can be left uncut, shoots longer than this should be cut back hard to leave about 2 cm (1 in.) of new growth each year. Thus the spur system is built up. This pruning is similar to that described for pruning the dessert gooseberry variety Leveller. Unfortunately this hard spur pruning must delay cropping and also reduce the total crop, although it increases individual berry size on the reduced number of strigs.

Red currants intended for mechanical harvesting by straddle type machines will have to be grown on a stool system to give a continuous hedge. All the shoots are allowed to develop full length and only thinned out and renewed by encouraging new growth from the base of the stool, a bit like a black currant. As there is no leader pruning the bushes may carry some fruit from the second year, natural spurs develop and the growth is not excessive. The shoots tend to become crowded and unfruitful at the base and as the fruit is carried on the tips the branches may droop sideways under a heavy crop. PYO growers are providing a trellis of three to four wires at about 50 cm (20 in.), 75 cm (30 in.), 100 cm (40 in.) and 150 cm (60 in.) above the ground, and weaving some of the shoots between these wires. Later growing side shoots are also woven in, and from time to time older branches that become too crowded are pruned away. Once the trellis is well covered with vertical shoots, side shoots can be pruned both in August and/or winter with a hedge cutter to leave about 25 cm of growth on both sides of the trellis.

Another method is to allow the unpruned new shoots to grow above two opposite supporting wires about 75 cm (30 in.) apart supported from cross bars nailed to posts about 40 cm (16 in.) to 50 cm (20 in.) above the ground. This support is just sufficient to keep the straggling shoots off the ground and allow the strigs to hang clear of the soil. Some hand labour is necessary to push the young shoots through the wirework of both systems. This is best done in summer, before picking, while the shoots are still supple. At this time any excessive growth can be snapped out and removed.

Control of weeds

Most red currants, like black currants and gooseberries, are now grown without cultivation, using herbicides to control weeds. The herbicides described for black currants (page 44) may be safely used for red currants.

Harvesting

Red currants are picked on the strig and should be uniformly coloured but not allowed to become over-ripe. Fruit intended for juice is usually picked into buckets and poured into standard wooden trays sent by the processor (page 64). Fruit for market may be picked direct into 2·5 kg (6 lb) fibreboard baskets, or into shallow wooden trays. Where the returns warrant the extra expense good quality fruit may be sold in 0·5 kg or 1 lb punnets protected within an outer container.

Pests and diseases

Red currants are not much troubled by pests and diseases. The most likely pests are aphids and capsids. The red currant blister aphid (*Cryptomyzus ribis*) frequently cause conspicuous red blistering of the leaves. Control measures are essentially similar to those suggested for aphids (page 54) and capsids (page 56) on black currants.

The red currant suffers from very few diseases and no routine control methods are necessary. Coral spot (page 94) may be seen on dead shoots and branches. Occasionally, as with black currants, bushes may be killed by *Armillaria* (page 60) and the black, boot-lace like rhizomorphs will be found on the dead roots. American gooseberry mildew (page 59) can attack red currants, causing the leaves to become mildewed, dry and curled upwards and to stop the growth of the shoot tip. If experience shows the disease to be troublesome preventative control methods will be needed but often attacks occur late after growth has finished. Bushes in nursery beds may need routine protection.

Certain virus diseases, but not reversion, can occur in stocks of red currants, causing vein banding and other symptoms comparable with those found in black currants and gooseberries. These diseases are not important at present but care should be taken to propagate only from apparently healthy bushes free from symptoms (page 94).

Bullfinches (Leaflet 234) damage the winter buds of red currants and like gooseberries red currants are much more likely to be attacked than black currants. As on gooseberries severe bulfinch damage not only reduces the crop but can spoil the shape of the bushes and result in long lengths of bare wood with no spurs or side branches. On small areas the protective nylon mesh cotton cording mentioned for gooseberries (page 95) may be worth trying, but overall reduction of damage is only possible with routine shooting and trapping. Sparrows (Leaflet 169) may attack the flowers in some seasons.

The ripe fruit of red currants is particularly attractive to blackbirds, song thrushes and starlings (Leaflet 208) and various scaring devices may have to be used just before and during the picking season to keep depredation to a minimum. Some of these devices may be subject to local bye-laws regarding use. Information on such local bye-laws and bird control can be obtained from Ministry Pest Control Officers located at the Ministry divisional offices.

Economics

The economics of the red currant can be calculated exactly as described for black currants (page 70). With the red currant it is even more important to ensure an outlet before planting any area of this crop, as the total area grown is so small and the demand very limited. Red currants also are not so popular as black currants on the wholesale market. Red currants are picked by hand at piecework rates about the same as are offered for black currants. There are no figures available for machine harvesting since this method is only being tested at present.

Red currants are slower to come into full crop compared with black currants, particularly if severely pruned. The life of the plantation is longer and it would not be unreasonable to allow 15–20 cropping years and to write off establishment costs over this longer period, allowing five years for the bushes to come into full crop.

Appendix 1

Other Ministry publications

REFERENCE BOOKS

RB35 Lime and Liming (90p)
RB95 Strawberries (£4·00)
RB107 Soils and Manures for Fruit (£1·89)
RB138 Irrigation (£2·75)
RB207 Apples (£3·25)
RB209 Fertiliser Recommendations (£3·00)
RB210 Organic Manures (£1·10)
RB426 Flowering Periods of Tree and Bush Fruits (£1·00)

The booklet *Approved Products for Farmers and Growers* listing materials in the Agricultural Chemicals Approval Scheme (insecticides/fungicides/herbicides) is revised annually.

The above publications are obtainable from HMSO PO Box 569 London SE1 9NH or can be ordered through any bookseller.

BOOKLETS

B2193 Lime and Fertiliser Recommendations: Fruit and Hops
B2196 Promising Varieties of Soft Fruits
B2198 Guidelines for the disposal of unwanted pesticides and containers on farms and holdings
B2263 Fruit Growers Guide to the Use of Chemical Sprays
B2264 Weed Control in Bush and Cane Fruits
B2272 Guidelines for applying crop protection chemicals
B2280 Windbreaks

LEAFLETS

11 Winter Moths
23 Coral Spot
30 Caterpillars on Currants and Gooseberry
57 Wingless Weevils
88 Scale Insects on Fruit Trees
115 Slugs and Snails
154 Capsid Bugs on Fruit

169 The House Sparrow
176 Currant and Gooseberry Aphids
208 The Starling
215 Gooseberries
226 Red Spider Mite on Outdoor Crops
234 The Bullfinch
259 Stem boring caterpillars of fruit plants
270 Soil Analysis for Advisory Purposes
273 Powdery Mildews of Gooseberry and Black Currant
277 Reversion Disease and Gall Mite of Black Currant
305 Bryobia Mites
320 Poultry Manure as a Fertiliser
435 Making the most of Farmyard Manure
441 Nitrogen Fertilisers
442 Phosphatic Fertilisers
443 Potassium and Sodium Fertilisers
444 Magnesian Lime and Magnesian Limestone
518 Lime in Horticulture
521 Red currants
543 Black currants
564 Soil-borne Virus Diseases of Fruit Plants
596 Magnesium fertilisers
617 Subsoiling
650 Black currant leaf spot

Appendix 2

Precautions

Whenever chemicals are used, follow the instructions given on the label on strength and frequency of application of the chemical and observe the recommended minimum intervals between application and planting or handling the crop. Read and follow carefully the Safety Precautions on the label. These should be as in the *Recommendations for the Safe Use of Chemical Compounds Used in Agriculture and Food Storage,* issued on individual chemicals and uses, by the Ministry of Agriculture, Fisheries and Food following clearance under the Pesticides Safety Precautions Scheme.

Use the chemicals mentioned in this Reference book with care, particularly those that are irritating to the skin, eyes, nose, and mouth such as:

    chlorothalonil, dinocap, mancozeb, manganese and zinc dithiocarbamate, thiram and zineb.

Do not allow any chemical to come in contact with skin or clothing.

Protection of consumers

To ensure that a harvested crop does not contain any harmful pesticide residue, follow the instructions given on the label on the strength and frequency of application of a pesticide and observe the minimum intervals between the last application and harvesting. For some pesticides mentioned in this Reference book minimum intervals are applicable as follows:

Pesticide	*Minimum interval between last application and harvesting*
azinphos-methyl plus demeton-S-methyl sulphone*	3 weeks
bupirimate	2 weeks
carbaryl	6 weeks (1 week for gooseberries)
chlorothalonil	3 days
chlorpyrifos	2 weeks
chlorthiamid	8 weeks

*Statutory Regulations apply to the use of these poisonous substances.
Some processors may request longer periods.

copper	3 weeks
demeton-S-methyl*	3 weeks
derris	1 day
dichlofluanid	3 weeks
dimethoate	1 week
dinocap	1 week
drazoxolon*	4 weeks
endosulfan*	6 weeks
fenitrothion	2 weeks
formothion	1 week
malathion	1 day
mancozeb	1 week
manganese and zinc dithiocarbamate . . .	1 week
methiocarb	1 week
oxydemeton-methyl*	3 weeks
phosphamidon*	3 weeks
quinomethionate	2 weeks
thiram	1 week
zineb	1 week

Dispose of empty containers safely, in accordance with the *Code of Practice for the Disposal of Unwanted Pesticides and Containers on Farms and Holdings.* Store new and part-used containers in a secure place. Do not contaminate ponds, ditches or waterways. (See Booklet 2198).

The Health and Safety at Work Act 1974 imposes general obligations on employers, the self-employed and employees that apply to work with any chemical. In addition the *Health and Safety (Agriculture) (Poisonous Substances) Regulations* lay down specific obligations in relation to the use of certain chemicals including:

    azinphos-methyl plus demeton-S-methyl sulphone
    demeton-S-methyl
    drazoxolon
    endosulfan
    oxydemeton-methyl
    phosphamidon

Users of any of these chemicals are strongly advised to read *The Safe Use of Poisonous Chemicals on the Farm* (available from the offices of the Health and Safety Executive) and *Guidelines for applying crop protection chemicals B2272* (available from MAFF offices).

Paraquat is available only to *bona fide* farmers or growers, who have to sign the poisons register on purchase.

Protection of bees

Do not spray when crops are in flower and cut down flowering weeds.

---

*Statutory Regulations apply to the use of these poisonous substances.
Some processors may request longer periods.

The drift of sprays on to neighbouring susceptible crops can cause considerable damage and must be avoided. See the *Code of Practice for Ground Spraying*.

READ THE LABEL

Copies of the leaflets and Codes of Practice referred to above are available free of charge from the address below or any Ministry regional or divisional office.

In order to choose a proprietary product the reader should consult the booklet of *Approved Products for Farmers and Growers* issued under the Agricultural Chemicals Approval Scheme of the Ministry and available as a priced publication from HMSO P.O. Box 569, London SE1 9NH or through booksellers. The label on the product gives detailed instructions on the dose to use, the dilution and the stage of growth of crop when spraying may be carried out. The container of an approved product bears the mark shown here. It is strongly recommended that Approved products only should be used.

Agricultural Chemicals
Approval Scheme

MINISTRY OF AGRICULTURE, FISHERIES AND FOOD
Publications), Tolcarne Drive, Pinner, Middlesex HA5 2DT

Index

All references are to black currant unless gooseberry or red currant is specified.

	Page		Page
ADAS soil sampling	35	Costs of establishment	74
Advisory leaflets	104	— annual	74
Analysis—soil	35	Cuttings	20
Analysis—leaf	38	— gooseberry	82
Annual weeds—lists of	51	— red currant	100
Aphids	54	— per area	29
— gooseberry	91	— planting distances	28
Approved chemical products	108	— spacing	23
Areas	2		
— gooseberry	77		
— red currant	97	Dichlobenil	47
Armillaria root rot	60	Die-back—gooseberry	93
		Diuron	46
Bed systems	27		
— gooseberry	85	Earwigs	57
Bees—protection of	107	Eelworm—leaf and bud	57
Bees—pollination by	19		
Big bud	55		
Bird damage—gooseberry	95		
— red currant	103	Fertilisers	33
Bird protection	78, 95	Flowers	19
Bullfinches—gooseberry	95	Frost—protection	8
Bush form—gooseberry	82	— radiation	7
— red currant	101	— wind	7
Bush spacing	26	Fruit rot	59
— gooseberry	83	FYM	34
— red currant	101		
		Gall mite	55
		Glyphosate	45
Capsid	56	Grey mould	7
— gooseberry	92	Grubbing	44
Certification schemes	20		
Chlorpropham	47		
Chlorthiamid	47		
Clearwing moth	56	Harvesting—hand	63
Consumer protection and sprays	106	— gooseberry	89
Coral spot—gooseberry	94	— mechanical	65

109

	Page
— gooseberry	90
Headlands	29
Herbicides	45
Herbicide programmes	48
Honey fungus	60
— gooseberry	94
Intercrop—gooseberry	78
Irrigation	40
— amounts	40
— equipment	10
Labour	4
— annual	72
— establishment	72
— harvesting	63, 76
— pruning	72
Leaf ash analyses	38
— midge	55
— spot	58
— spot—gooseberry	94
Leveller picking	90
— pruning	89
Lime	36
Magnesium	37
Manuring	33
— gooseberry	85
— red currant	101
MCPB	48
Mechanical harvesting—costs	75
— gooseberry	90, 96
— destructive	65
— labour	75
— outputs	75
— plucker belt	66
— pruning for	43
— semi-mounted	67
— straddle	67
Midge	55
Mildew	59
— gooseberry	92
Minor elements	38
Mulching	35
— gooseberry	87
Multiple row plant	27
Nitrogen	36
Number of bushes per area	27
Nursery bushes	24

	Page
Organic matter	34
Outlets	3
Paraquat	46
Perennial weeds—list of	52
Phosphorus	36
Picking rates	63
— gooseberry	90
Planting	32
— depth	32
— distances	27
— gooseberry	84
Pollination	81
— gooseberry	80
Potassium	37
Processing—gooseberry	77
— outlets	3
— prices	71
— red currants	97
— and spraying	54
Propyzamide	47
Pruning—after planting	42
— disposal of	44
— red currant	101
— older bushes	43
— younger bushes	42
— older gooseberry bushes	88
— younger gooseberry bushes	87
— tools	41
Red spider mite	56
Reversion virus disease	60
Row length	29
Rust	60
— gooseberry	94
Sawfly	56
— gooseberry	91
Simazine	46
Slugs	57
Snails	57
Soils	11
Soil sampling	35
Spray chemicals—mixing	53
— programmes	54
Sprayers—air-carrier	53
— ground	50
— high volume	52
— precautions	106
Sprinklers	9
Stools	1

	Page
— red currants	101
Straw mulching	36
Subsoiling	30
Sucker shoots	88
Trays—market	65
— processing	64
Varieties	13
— gooseberry	80
— red currant	98
Virus	60

	Page
— gooseberry	94
Vitamin C	14
Wind damage—gooseberry	80, 83
— red currant	98
Windbreaks	6
— gooseberry	79
Yields	3
— gooseberry	78
— red currant	97